GOVERNING RENEWABLE NATURAL RESOURCES

In one volume, this book brings together a diversity of approaches, theory and frameworks that can be used to analyse the governance of renewable natural resources.

Renewable natural resources are under pressure, with over-exploitation and degradation raising concern globally. Understanding governance systems and practice is essential for developing effective and fair solutions. This book introduces readers to key concepts and issues concerned with the governance of renewable natural resources and illustrates the diversity of approaches, theories and frameworks that have been used to analyse governance systems and practice. Each chapter provides an introduction to an area of literature and theory and demonstrates application through a case study. The book covers a range of geographical locations, with a focus on low- and middle-income countries, and several types of natural resources. The approaches and theories introduced include common property theory, political ecology, institutional analysis, the social-ecological systems framework and social network analysis. Findings from across the chapters support an analytical focus on institutions and local context and a practical focus on diverse, flexible and inclusive governance solutions.

The book serves as an essential introduction to the governance of renewable natural resources for students, researchers and practitioners.

Fiona Nunan is Professor of Environment and Development in the International Development Department, University of Birmingham, UK.

Earthscan Studies in Natural Resource Management

Sustainability Certification Schemes in the Agricultural and Natural Resource Sectors
Outcomes for Society and the Environment
Edited by Melissa Vogt

Policy and Governance in the Water-Energy-Food Nexus
A Relational Equity Approach
Edited by Anastasia Koulouri and Nikolai Mouraviev

UNESCO Biosphere Reserves
Supporting Biocultural Diversity, Sustainability and Society
Edited by Maureen G. Reed and Martin F. Price

Social Movements Contesting Natural Resource Development
Edited by John F. Devlin

Benefit-sharing in Environmental Governance
Local Experiences of a Global Concept
Louisa Parks

Governing Renewable Natural Resources
Theories and Frameworks
Edited by Fiona Nunan

Interdisciplinary Collaboration for Water Diplomacy
A Principled and Pragmatic Approach
Edited by Shafiqul Islam and Kevin M. Smith

For more information on books in the Earthscan Studies in Natural Resource Management series, please visit the series page on the Routledge website: *http://www.routledge.com/books/series/ECNRM/*

GOVERNING RENEWABLE NATURAL RESOURCES

Theories and Frameworks

Edited by Fiona Nunan

Routledge
Taylor & Francis Group
LONDON AND NEW YORK

earthscan
from Routledge

First published 2020
by Routledge
2 Park Square, Milton Park, Abingdon, Oxon OX14 4RN

and by Routledge
52 Vanderbilt Avenue, New York, NY 10017

Routledge is an imprint of the Taylor & Francis Group, an informa business

British Library Cataloguing-in-Publication Data
Names: Nunan, Fiona, editor.
Title: Governing renewable natural resources: theories and frameworks / Edited by Fiona Nunan.
Description: Abingdon, Oxon; New York, NY: Routledge, 2020. | Series: Earthscan studies in natural resource management | Includes bibliographical references and index.
Identifiers: LCCN 2019034538 (print) | LCCN 2019034539 (ebook) | ISBN 9780367146696 (hardback) | ISBN 9780367146702 (paperback) | ISBN 9780429053009 (ebook)
Subjects: LCSH: Renewable natural resources—Government policy. | Renewable natural resources—Management. | Environmental policy.
Classification: LCC HC85 .G70 2020 (print) | LCC HC85 (ebook) | DDC 333.7—dc23
LC record available at https://lccn.loc.gov/2019034538
LC ebook record available at https://lccn.loc.gov/2019034539

Library of Congress Cataloging-in-Publication Data
A catalog record has been requested for this book

ISBN: 978-0-367-14669-6 (hbk)
ISBN: 978-0-367-14670-2 (pbk)
ISBN: 978-0-429-05300-9 (ebk)

Typeset in Bembo
by codeMantra

CONTENTS

ILLUSTRATIONS

Figures

Tables

Boxes

CONTRIBUTORS

Shankar Aswani is Professor of Anthropology and Ichthyology & Fisheries Science at Rhodes University, South Africa. His research has focused on a diversity of subjects including property rights and common pool resources, marine indigenous ecological knowledge/ethnobiology, vulnerability and resilience of coastal communities, human behavioural ecology of fishing, economic anthropology, tourism, ethnohistory and applied anthropology. He is involved in designing and establishing marine conservation programs in Oceania, currently heading projects on coastal vulnerability in Tanzania, Angola, Mozambique, Madagascar and South Africa.

Clare Barnes is an interdisciplinary lecturer in Sustainable Livelihoods in the School of Geosciences at the University of Edinburgh. Her research interests are in natural resource governance in the Global South. Research themes include land rights, policy change, rural livelihoods and the roles of civil society organisations. Clare's PhD, from Utrecht University in The Netherlands, was on community forest governance in India.

Örjan Bodin is an associate professor at the Stockholm Resilience Centre at Stockholm University. He studies different challenges and opportunities in managing natural resources, and much of his research bridges the natural and social sciences.

Papa Faye holds PhDs in Sociology and Social Anthropology from Université Cheikh Anta Diop of Dakar and from the Institute of Social Anthropology at University of Bern in Switzerland. He has been conducting research on natural resource governance with national and international research institutions since 2004. Papa is co-founder and, since 2017, Executive Secretary of an action-research centre based in Dakar, known as the Centre d'Action pour le Développement et la Recherche (see *www.cadre.sn*).

Mikkel Funder is a senior researcher at the Danish Institute for International Studies. He works on the governance of natural resources, environment and climate change in the context of development. He is particularly interested in everyday governance and institutional change, and how actors shape and respond to environmental agendas and interventions in practice.

Ingela Jansson is a PhD candidate at the Swedish University of Agricultural Sciences. Ingela has worked in the Serengeti Ecosystem since 2006, initially as a research assistant for the Serengeti Lion Project. Since 2011, she has been based full-time in Ngorongoro Conservation Area (NCA), as co-founder and director of KopeLion, an organization to foster human–lion coexistence in NCA through participatory human-wildlife conflict mitigation, lion monitoring and lion protection.

Prakash Kashwan is an associate professor in the Department of Political Science, University of Connecticut. He is the author of *Democracy in the Woods: Environmental Conservation and Social Justice in India, Tanzania, and Mexico* (Oxford University Press, 2017). His research and scholarship focusses on political economy of institutions, environmental policy and politics, global climate governance and climate justice. He is an associate editor of the journal Progress in Development Studies.

María Mancilla García is a researcher at the Stockholm Resilience Centre in Sweden. She works on relational approaches to the study of social-ecological systems. Her work ranges from applying process-relational philosophy to the analysis of social-ecological systems as complex adaptive systems to conducting empirical studies on water governance and fisheries governance in Latin America and Africa.

Fiona Nunan is Professor of Environment and Development in the International Development Department, University of Birmingham, with over 25 years' experience in environment and development. Her research interests focus on the governance of renewable natural resources, particularly inland fisheries and coastal ecosystems, and she is the author of *Understanding Poverty and the Environment: Analytical frameworks and approaches* and editor of *Making Climate Compatible Development Happen*.

Stefan Partelow is an environmental social scientist at the Leibniz Centre for Tropical Marine Research (ZMT) in Germany. His research examines governance and collective action processes in marine and coastal systems with a focus on small-scale fisheries, aquaculture and tourism. This work broadly aims to advance knowledge on the concepts of governance, social-ecological systems and sustainability.

Adam Pekor is an attorney and chair of the board of directors of KopeLion, a community-based lion conservation organization located in Tanzania's Ngorongoro Conservation Area. Adam focuses on incentive-based measures to mitigate human–wildlife conflict and promote local participation in conservation. He holds a law degree from Georgetown University and a master's degree in conservation biology from Columbia University.

Dennis Rentsch, PhD, University of Minnesota, has worked in the Serengeti-Ngorongoro ecosystem for over 15 years, managing community-based conservation initiatives and conducting applied research with communities living within the ecosystem. From 2008–2016, he led the Serengeti Ecosystem Management office for the Frankfurt Zoological Society, and in 2016 he joined Lincoln Park Zoo as Serengeti Research Scientist, directing the Zoo's research in the ecosystem to focus on zones of interaction between humans and wildlife to reduce the negative impacts on people and wildlife.

Camilla Sandström, Professor in Political Science, Umeå University, is an expert on governance and management of natural resources, especially forests, wildlife and biological diversity. She is experienced in working with issues about socioecological resilience and methods to handle conflicts of interest concerning natural resources and land use. She runs the Environmental Policy and Governance group at Umeå University and co-leads the interdisciplinary research group Society and Wildlife.

William Ole Seki, BSc in Development Studies from Kampala International University, is a traditional Maasai from a Kisongo sect of the Maasai tribe. A born conservationist and human rights lobbyist. Highly experienced in pastoralism livelihoods and the indigenous knowledge that propels it. A devout believer in participatory development through genuine policy formulation, implementation and analysis.

Göran Spong runs the Molecular Ecology Group at the Swedish University of Agricultural Sciences and has a broad interest in behavioral ecology and population biology of terrestrial vertebrates. Many of the species his research group study, like the lion, are of economic or conservation concern. Genomics and non-invasive genetic approaches are typically used to address elusive or cryptic processes. His work ranges from fundamental and applied research to commissions in management and forensics.

Caroline Upton is a human geographer, based in the School of Geography, Geology and the Environment, University of Leicester. She has worked with pastoralist communities in Mongolia for over 15 years. Her work focuses on issues of environmental governance and justice, climate change, conservation and livelihoods, often deploying a political ecology approach.

1

INTRODUCTION

Governing renewable natural resources

Fiona Nunan

Introduction

The status of the world's renewable natural resources remains under severe pressure. Marine fish stocks continue to decline (FAO, 2018a); deforestation and land use change continue to contribute to loss of habitat, soil erosion and climate change (FAO, 2018b); and biodiversity loss continues at an alarming rate (Secretariat of the Convention on Biological Diversity, 2014). Responding to these worrying trends requires not only an understanding of the natural resource systems themselves but of who is exploiting them and for whose benefit. Managing natural resources then is not so much about managing the resources themselves but about managing people. In managing people, decisions have to be made about what resources are being managed for and whether products can be extracted, by whom, how much, when and from where. These decisions are made within the sphere of governance. Analyzing such governance can generate understanding about who is making these decisions, why those actors are involved in decision-making and who is benefiting from the decisions.

This book provides an introduction to, and a reflection on, a diverse range of theories, frameworks and approaches that have been drawn on to investigate and understand governance systems and outcomes associated with renewable natural resources. There is no one theory that explains how governance works or what it should look like. There is no single framework or analytical approach that can generate data to answer all questions that might be asked of governance, such as who is involved, why are specific decisions made and is governance leading to sustainability of ecosystems and improved livelihoods of people who depend on natural resources. This book brings together just some of the areas of theory, types of frameworks and key approaches that have been used to analyze governance. In doing so, it provides a guide to students, researchers and practitioners to the kinds of questions that can be asked of governance and that matter for sustainability and human development, to the tools and methods that can be used to generate and analyze data and to the kinds of findings that provide insights into how governance is working, why and with what implications. The book focuses on examples from low- and middle-income countries, with case studies taken from countries including India, Mongolia, Brazil, Senegal, Tanzania and Zambia, whilst several chapters take a global perspective. The focus on low- and middle-income countries encourages attention to distributional concerns, given the level of livelihood dependence on renewable natural resources in many of these countries.

Each chapter introduces, explains and briefly reviews one approach, area of theory or a framework that has been, or could be, used to investigate governance. Following this, the theory, framework or approach is applied to a specific case study to illustrate what application might involve and what kinds of findings may emerge. The case studies are mainly local in character, though some are national or take a global perspective.

The purpose of this chapter is to introduce the concept of governance, the main approaches that have been taken to governing renewable natural resources and the key challenges and issues that have been identified through research as compromising the potential for governance to deliver on sustainability and human development. This provides the reader with a grounding in which to appreciate the detailed analyses of governance systems, processes and outcomes provided in the following chapters.

What is governance?

Governance is broadly viewed as referring to 'the formation and stewardship of the formal and informal rules that regulate the public realm, the arena in which state as well as economic and societal actors interact to make decisions' (Hyden et al., 2004: 16). With specific reference to natural resources, Campese (2016: 7) defines natural resource governance as,

> the norms, institutions, and processes that determine how power and responsibilities over natural resources are exercised, how decisions are taken and how citizens – including women, men, youth, indigenous peoples and local communities – secure access to, participate in, and are impacted by the management of natural resources.

This second definition does not stress the role of the state in governance but refers to the 'norms, institutions and processes' that influence decision-making on how natural resources are managed and who has access to natural resources and for what purposes. Governance, then, does not only encompass structures, systems and actors but also processes, norms and institutions, with institutions understood as 'rules of the game' (North, 1990: 3). The distribution and manifestation of power is a key concern within governance as these have consequences for which interests are represented, whose voices are heard and with what implications for ecosystem health and livelihoods. Given these definitions and understanding of governance, it is evident that there may be many actors, structures and institutions involved and so understanding a governance system, its practice and performance, may be a complex exercise.

As governance is concerned with who is making decisions, why and how, a lot of attention has been paid to how transparent and accountable that decision-making is. This has been demonstrated through programmes and measurements of 'good governance' and the development and application of 'governance principles', as well as a suite of governance indices and indicators. The pursuit of 'good governance' is associated with the response to perceptions of the failure of countries in sub-Saharan Africa to develop as expected in the 1980s, with blame for this failure attributed to 'bad governance' (World Bank, 1989). The model of 'good governance' was founded on Anglo-American liberal-democratic states and led to the development of lists of 'ingredients' of good governance. From a review of such lists developed by donor agencies, Gisselquist (2012) identified seven core components as 'democracy and representation, human rights, the rule of law, efficient and effective public management, transparency and accountability, developmentalist objectives and a varying range of particular political and economic policies, programmes, and institutions' (2012: 2).

Whilst such lists are not universally supported, with some commentators viewing them as unrealistic and unhelpful, such as Grindle (2004, 2007) arguing for 'good enough governance' rather than long lists of demands, indicators and principles of good governance persist. Within natural resource governance, several lists of governance principles have been developed, with Lockwood et al. (2010: 997) suggesting that governance principles 'provide normative guidance for the establishment of good-practice multilevel NRM [natural resource management] governance'. Lockwood et al.'s (2010) set of eight principles consist of legitimacy, transparency, accountability, inclusiveness, fairness, integration, capability and adaptability. A similar list is provided by Springer (2016) in a document setting out a potential Natural Resource Governance Framework for the International Union for Conservation of Nature (IUCN). This list has 12 principles: inclusive decision-making, recognition and respect for legitimate tenure rights, devolution, diversity of cultures and knowledge, strategic vision, empowerment, coordination and coherence, sustainable resources and livelihoods, social and environmental accountability, protection of the vulnerable, rule of law and access to justice. Both lists illustrate the desirability of governance to be transparent, inclusive, accountable and participatory. Such principles are reflected in some of the challenges found from research and practice in natural resource governance and are the subject of some of the analyses reported on in this book.

Why renewable natural resources

As stated at the beginning of this chapter, this book is concerned with renewable natural resources, particularly in low- and middle-income countries. There are several reasons for this focus. First, many renewable natural resources are considered to be 'common pool resources', that is, a type of economic good that has high costs of exclusion (it is difficult to stop other people from using the resource) and high levels of subtractability (the consumption by one user will reduce the quantity available for others) (Ostrom et al., 2002). Of course, renewable natural resources should be able to 'renew' themselves, that is, replenish their stocks, as opposed to non-renewable natural resources, such as oil, natural gas and minerals, which do not. However, this takes time and appropriate conditions, and so extraction may well reduce the amount available to others, whether this is fish, grass for grazing livestock or timber. These characteristics of common pool resources mean that decision-making is needed for renewable natural resources to remain 'renewable' (i.e. sustainable management) and to allow or restrict access and extraction of products.

Second, many renewable natural resources provide multiple benefits – that is, many ecosystem services. This results in many people benefiting from resources and potentially multiple management objectives being pursued, which may include employment creation, revenue generation, income generation and good ecosystem health. Objectives which are in conflict with each other could exist, particularly where objectives come from different actors, including different parts of government. In a mangrove forest, for example, management objectives may come from ministries associated with forestry, coastal protection, fisheries and climate change. The existence of multiple objectives, actors and structures may result in a complex system of governance, one that it is not necessarily well planned and coordinated but is fragmented and lacking in coordination (Nunan, 2018). The scale and complexity of both natural and social systems and their interactions may present further challenges to coordination, crossing administrative and national boundaries. Further, it must be noted that it is not always, perhaps rarely, possible to achieve multiple, and at times contrasting, objectives. This means that identifying and assessing trade-offs in decision-making is often a necessary part of governance (Nunan et al., 2018).

A third reason for focusing on renewable natural resources, and particularly in relation to low- and middle-income countries, is because of the direct role they play in many people's livelihoods, in addition to indirect provision of ecosystem services such as carbon sequestration and coastal protection. This means that the question of who has access to natural resources, and which natural resource products, is in part a question of governance. Many factors can influence access, including property rights, social norms and institutions (Nunan, 2015). Governance can influence access through decision-making on rules and regulations and who is involved in such decision-making matters.

This suggests that the systems and practice of governance are themselves affected by a range of social norms and institutions, with the distribution of power influencing who is involved in making decisions, what those decisions are and whom they benefit.

Approaches to governing renewable natural resources

There are many approaches taken to governing renewable natural resources. These are not mutually exclusive and may exist within the same system, with or without integration, coordination or cooperation. Within any governance system, there is usually some element of state-centred regulation and enforcement, with central government leading on policy, legislation and enforcement. Until the 1980s, when decentralization of government authority and functions and participation of resource users in management was widely adopted, this approach dominated. State-led governance is characterized by a 'fences-and-fines' approach, where resource users are kept out of protected areas, for example, and stopped from extracting any resources or grazing their livestock in fenced areas (Adams and Hulme, 2001). This approach remains significant and influential, even in cases where seemingly communities, lower levels of government and the private sector should be dominant in governance arrangements. Benjaminsen and Svarstad (2010), for example, argue that despite the rhetoric of win-win discourse on protected areas, where communities and wildlife are purported to benefit, 'fortress conservation' prevails in Tanzania and South Africa, with the poor bearing the cost of wildlife conservation.

Customary systems are also of significance in governing renewable natural resources. Traditional or customary structures and rules have long existed and played a role in determining how natural resources are used and by whom. Customary systems may or may not be recognized in national law and may or may not work with other systems or necessarily be seen as delivering on fair and equitable processes and outcomes. Examples of customary systems that form part of a governance system are given by Karnad (2017), who details how customary law within fisheries in Maharashtra state, India, protects common property rights in contrast to ineffective state legislation, and Tadie and Fischer (2017), who demonstrate how customary institutions use negotiation and reciprocity instead of rules and regulations to govern access to resources by pastoralist communities in lower Omo, Ethiopia.

Different approaches to decentralization have been adopted within natural resource governance, all involving a transfer of authority from central to local government. Decentralization may involve the transfer of power and resources to locally representative bodies, with elected members accountable to their constituents. Alternatively, there may be a deconcentration of power from the centre to lower levels, with branch offices reporting back to the centre. Finally, devolution of power refers to the transfer of power from central government to a lower level, which may include community-level organizations (Nunan, 2015). There can, then, be confusion about what is meant by 'decentralization' within literature with some sources including devolution to communities and others suggesting that the term refers only to decentralization to democratically elected bodies. Critiques of the practice of decentralization within natural resource governance include concern that democratically elected local governments are bypassed, with forest departments, for example, devolving power to forest officers and community forest management structures rather than local government, thereby reducing the potential for accountability to local communities (Ece et al., 2017; Ribot et al., 2010). A further critique is the limited power that is devolved in practice, with Poteete and Ribot referring to tactics used to maintain control as 'repertoires of domination' to describe 'the myriad tactics government officials and nonstate actors use to limit meaningful shifts of authority associated with decentralization' (2011: 440).

Devolution of power and responsibilities to groups of resource users has led to the development of participatory or community-based management approaches. These differ from customary or traditional systems by being often developed through top-down approaches led by an external actor, whether government, donor-funded projects or non-governmental organizations (NGOs). The generic term community-based natural resource management (CBNRM) refers to a plethora of approaches and terms, such as community-based conservation and community forest management, and may also be used to refer to more collaborative approaches, where communities work with other actors, particularly government, in managing a resource, such as Joint Forest Management and fisheries co-management. Such approaches have been seen as a solution to the limited resources and capacity of government, as well as a way of empowering local people and enabling their voice and preferences to shape decision-making. CBNRM is therefore often portrayed as being capable of delivering on 'win-win' outcomes, where the environment is protected and enhanced and local livelihoods are improved. In practice, a number of challenges have been widely experienced, including capture of decision-making structures and resources by local elite, more marginalized actors not being well represented in decision-making structures and limited, if any, improvements in the environment and livelihoods experienced (Anderson and Mehta, 2013; Galvin et al., 2018). A further critique of CBNRM-related initiatives is the way that communities are often portrayed and understood, with inadequate acknowledgement of differences within place-based communities, including those in terms of power, social status and preferences (Agrawal and Gibson, 1999; Ojha et al., 2016).

A final key form of governance approach is where a market-type mechanism influences access to natural resources and markets, such as certification schemes and Payment for Ecosystem Services (PES) schemes. Certification schemes aim to facilitate tracing of resources to their origin and provide assurances of origin and quality, whereas PES schemes provide incentives, or compensation, for conservation. Both can involve other forms of governance approaches, including community-based approaches where communities rather than individuals are involved, though they are particularly noted for bringing in private sector actors to the governance arena (Cashore et al., 2004; Matzdorf et al., 2013). Central government often still plays a strong role, influencing the uptake and direction of such schemes (Giessen et al., 2016; Matzdorf et al., 2013).

Factors that influence the governance of renewable natural resources

Thinking about governance with respect to renewable natural resources is informed by many areas of theory and literature, including broader literature on governance (e.g. Kooiman, 2003), literature on governance principles (e.g. see Lockwood et al., 2010) and common property theory, associated, in particular, with the work of Elinor Ostrom (e.g. Ostrom, 1990). Ostrom's work has been influential as she identified a set of conditions, or 'design principles', under which common pool resources may be successfully governed collectively. These are shown in Box 1.1. Common property theory offered evidence and rationale for how people do, and can, work together in managing natural resources, suggesting that common property regimes are a viable alternative to state and private property regimes.

BOX 1.1 OSTROM'S DESIGN PRINCIPLES FOR EFFECTIVE COMMON PROPERTY REGIMES

1. There are clearly defined boundaries for the common pool resource being governed and who can benefit from the resource.
2. Rules governing use or provision of the resource should be appropriate to local conditions and context.
3. There should be collective-choice arrangements in place, where those affected by operational rules can participate in making and reviewing those rules.
4. Monitoring of rules and use should be undertaken by users or be accountable to the users.
5. Graduated sanctions should be agreed and enforced on infringements, depending on factors such as the type of offence and the number of times the offence has occurred.
6. Conflict resolution mechanisms should be in place so that resolution can take place quickly and at low cost.
7. Recognition of legitimacy of those using the resource to make their own rules is essential.
8. Nested enterprises in multiple layers may be needed for common pool resources that are part of larger systems.

Source: Adapted from Ostrom (1990: 90)

These principles have informed governance approaches such as CBNRM, with Jones (2010) attributing guidance for the formation of Conservation Conservancies in Namibia to Ostrom's principles. Ostrom herself noted, however, that these design principles are no guarantee for success and emphasized that there is no one approach to governance that will work in all circumstances, observing that there are 'no panaceas' (Ostrom, 2007). She advised that

> We should stop striving for simple answers to complex problems – no panaceas. Instead, we need to recognize and understand the complexity to develop diagnostic methods to identify combinations of variables that affect the incentives and actions of actors under diverse governance systems.
>
> *(Ostrom, 2007: 15181)*

Research and practice have tested and further developed this list of principles, with scholars identifying further factors that are essential for effective commons governance (see Agrawal, 2001, for example) and suggesting that this list is most appropriately used in a diagnostic, rather than blueprint, approach (Cox et al., 2010), confirming Ostrom's own view of there being no panacea. Cox et al. (2010) review applications of Ostrom's principles and confirm their relevance though acknowledge that there is scope for additional aspects and perspectives to be reflected in explanatory variables, particularly biophysical, cultural and political ecology. Agrawal (2003) also identifies missing factors, including those from the wider social and political-institutional environment, with a greater focus on power dynamics within communities and in the governance of common pool resources called for.

Literature on factors that shape and influence governance complements and supplements literature on approaches to governance and literature on governance principles. These areas of literature are each examples of analyses of governance systems, performance and outcomes and so illustrate the diversity of approaches that can be taken to researching governance.

Challenges to effective governance of renewable natural resources

Another way of looking at factors which influence the performance and outcomes of governance is by identifying and examining challenges experienced in the practice of governance. Not all challenges that have been identified in the vast literature on governing renewable natural resources can be reviewed here. Instead, the section focuses on key areas of challenges that have been identified, clustering several issues within sections on scale and complexity, and power and participation.

Scale and complexity

It was noted earlier in the chapter that governance involves not only structures, systems and actors but also processes, norms and institutions. This suggests that the arena of governance may often be complex, with multiple dimensions, components and interactions. In addition, one of the rationales for focusing on renewable natural resources was given as the multiple ecosystem services provided, often resulting in there being multiple parts of government involved as well as other actors, including resource users, the private sector, NGOs and regional and international organizations. These components and range of ecosystem services can be found at multiple administrative and spatial and temporal levels and scales. For example, resource users may make decisions at the local level, and, at the same time, benefits may be experienced globally from carbon sequestration. Actors at multiple levels and scales may form a governance system, from international certification schemes to local, place-based communities. In addition, some natural resources may cross administrative boundaries, within or between countries, leading to a need for cooperation and coordination.

The complexity of governance system has been portrayed, explained and analyzed in multiple ways, for example as multilevel governance, polycentricity and network governance (Nunan, 2018). Multilevel governance reflects the multiple levels (usually administrative) at which actors involved in governance are located and highlights the existence of, and need for, horizontal (cross-level) and vertical (between levels) interactions (Armitage, 2008; Mwangi and Wardell, 2012). Polycentricity also recognizes multiplicity but does so in terms of there being multiple centres of decision-making, where 'decision making authority is distributed in a nested hierarchy and does not reside at one single level' (Pahl-Wostl, 2009: 357). A network perspective on governance recognizes that actors may reach out to other actors across levels, not necessarily going through hierarchical systems, endeavouring to share information, make useful connections and work with other stakeholders to effectively govern natural resources (Carlsson and Sandström, 2008).

Challenges associated with multilevel governance include the difficulty of coordination and cooperation. Actors involved may not be incentivized to coordinate their priorities, plans and activities or may lack the resources and capacity to work collaboratively with other actors, including other parts and levels of government. Taking a multilevel perspective encourages attention to the existence of, and potential for, vertical and horizontal interaction, while polycentricity suggests that a nested hierarchy and coordination of activities will lead to effective governance. Network governance also responds to the challenge of cooperation and coordination in a complex governance system, with actors forming networks through their interactions that may complement or bypass formal hierarchical structures. Additional challenges associated with multilevel governance include a lack of transparency and accountability in decision-making and a lack of representation, particularly from the local level (Poteete, 2012; Termeer et al., 2010).

Complexity within governance is also seen in the response to uncertainty, inadequate information and change in social-ecological systems. Such a response is encapsulated within adaptive governance, in which governance systems have capacity to respond to new information and to change, generate and share knowledge, feeding information back into decision-making, and encouraging attention to processes rather than structures for greater effectiveness and for building adaptive capacity (Chaffin et al., 2014; Folke et al., 2005; Williams, 2011). Adaptive co-management is a form of adaptive governance, where collaboration and power-sharing enable the development of responsive processes and information sharing (Armitage et al., 2007).

Power and participation

A second key area in which several topics and issues are clustered is 'power and participation'. Decentralized, community-based and collaborative forms of governance facilitate participation of a range of actors, particularly community-based resource users, in governance systems. Effective participation, however, is far from easy to achieve in practice. This is partly because the very word 'participation' is hard to pin down and can be interpreted in multiple ways, from minimal consultation to handing over power to resource users (Cornwall, 2008). Participatory approaches within natural resource governance may therefore involve varying degrees of participation, with some programmes claiming participation when only brief consultation has taken place (Shackleton et al., 2010).

An important critique of participatory approaches is that they do not always take into consideration the existing power structures and politics. Any approach that is not informed by understanding of power differentials and political dynamics may risk further marginalizing already marginalized people yet claim that participation has taken place (Cornwall, 2008). Participation of women and marginalized groups in CBNRM structures may be more limited than that of men, particularly wealthier men of a high social status. There are believed to be far fewer women involved in fisheries management committees, though the evidence is limited (Alonso-Población and Siar, 2018). Within forest management, most evidence comes from Nepal and India, which although positive about women's participation in user committees does not imply that the situation is necessarily as positive in other parts of the world (Leisher et al., 2016). More marginalized groups, such as lower caste groups, may also find it difficult to be represented on user group, with Adhikaria and Di Falco (2009) finding, for example, that people of a lower caste in Nepal were less likely to be on a forest user group than men from higher castes.

Differences in the experience of participation in natural resource governance reflect differences in power. Power is manifested in multiple ways in natural resource governance, including the degree and nature of power devolved by central government to lower levels of government and to local resource users and power relations within communities and households, influencing how resources are accessed and used and who participates in decision-making over natural resources. It has been consistently shown that governments are reluctant to devolve adequate power for community-based structures to play an effective role in natural resource governance (e.g. see Baynes et al., 2016; Ribot et al., 2010), and so power relations between community groups and government officials shape and inform governance processes and performance.

A prevalent example of power relations at work within CBNRM is 'elite capture'. This refers to situations where actors with high social status, power and, often, greater wealth dominate management systems and decision-making and work to ensure that their interests are protected or promoted. They may capture key positions on committees for themselves and individuals willing to support them, misappropriate funds and control decision-making (Muyengwa et al., 2014). There is some evidence that, over time, the effects of elite capture can be reduced through non-elites challenging power structures (Lund and Saito-Jensen, 2013) and external agencies, such as NGOs, supporting communities in establishing CBNRM, which has been shown to reduce the negative effect and scale of elite capture (Persha and Andersson, 2014).

Analytical frameworks

A number of analytical frameworks exist that bring together elements of the dimensions, components and practice of governance reviewed in this chapter. Nunan (2015) reviews examples of such frameworks within the context of reviewing frameworks that enable analysis of relationships between poverty and the environment. A comparison of the four frameworks reviewed in Nunan (2015) is provided in Table 1.1. As can be seen from the table, the first framework focuses on decentralization, identifying three dimensions that should be analyzed to understand how decentralization of the governance of renewable natural resources is working. These dimensions are actors, power and accountability. In the second framework, Cleaver and Hamada (2010) present a modified version of a framework provided by Franks and Cleaver (2007), in which mechanisms are used by actors and agents to mediate relationships between resources and outcomes for people and ecosystems. In the third approach of interactive governance and its associated governability assessment, governing systems, the system to be governed and governing interactions are identified and analyzed. From this the 'governability' of a system can be assessed in terms of the diversity, complexity, dynamics and scale of the systems and their interactions. Finally, in the fourth approach to analysis, Batterbury and Fernando (2006) identified five dimensions that should be examined in an analysis of natural resource governance.

TABLE 1.1 Comparison of approaches to the analysis of governance

Decentralization (Agrawal and Ribot, 1999)	*Water governance and poverty (Cleaver and Hamada, 2010)*	*Interactive governance and governability (Kooiman et al., 2008)*	*Analysis of governance (Batterbury and Fernando, 2006)*
Actors	Actors and agents	Governing system	Historical context needed
Power	Resources	System to be governed: human/ societal and natural systems	Political economy of governance
Accountability	Mechanisms	Governing interactions	Scale and levels
	Outcomes for people and ecosystems	Analysis in terms of: • Diversity • Complexity • Dynamics • Scale	Rules in use
	Processes of management and practice		Political ecology of governance

Source: Adapted from Nunan (2015: 173).

From Table 1.1, it can be seen that there are similarities and differences in approaches taken to analyzing the governance of renewable natural resources. In terms of similarities, there are commonalities in a focus on identifying who is involved in governing and how the governance system is working in practice and with what outcomes. They differ in terms of whether the wider political economic context is investigated, or acknowledged, and whether an historical approach is taken. These frameworks reflect the need to understand the different dimensions, components and practice reviewed in this chapter. This includes the different approaches taken to governance, the principles and factors that inform governance approaches and are used to appraise governance practice and outcomes, and the challenges experienced in practice and outcomes due to scale, complexity and power relations. These dimensions and components are reflected in the frameworks, approaches and theories introduced and drawn on in the following chapters.

Structure of the book

This chapter has set the scene for the book by introducing key concepts, approaches and issues related to the governance of renewable natural resources. In the following chapter (Chapter 2), Kashwan builds on this by developing an overarching conceptual framework to analyze key elements of governance, informed by the political economy of institutions perspective. He argues that the framework addresses key gaps in existing frameworks, in relation to the specification of goals, how stakeholders are represented, institutional arrangements for the monitoring and enforcement of regulations and the state demonstrating credible commitment and recognition of the distribution of power. This meta-framework is then applied to three examples of frameworks for assessing natural resource governance to demonstrate its utility.

Chapter 3 introduces social network analysis (SNA) as a method to map and analyze social relations. Mancilla Garcia and Bodin demonstrate how SNA can inform analyses of participation in governance, drawing on an example of a water basin committee in Brazil. Through this analysis, they illustrate how a network-based understanding of participatory forums can help identify and address challenges to inclusive governance institutions.

The social-ecological systems framework (SESF) is introduced and applied in Chapter 4. Partelow provides a brief history of scholarship behind the framework and then introduces and applies the framework itself, noting how the framework brings systems thinking to the analysis of natural resource governance. The case study is of a co-management system in Brazil, the Brazilian Extractive Reserve (RESEX) program, which gave common property and co-management rights back to local and indigenous natural resource users. Partelow demonstrates how the SESF can enable identification of the key influencing variables in a situation and facilitate comparison between cases.

Chapter 5 introduces an alternative approach to the analysis of institutions within the context of natural resource governance. This is critical institutionalism, an approach which recognizes the influence and role of institutions embedded in social and cultural life in natural resource governance. This approach presents a shift in thinking away from institutions being created specifically for the purpose of natural resource governance towards those that have emerged within societies. The approach further recognizes that institutions may change over time and interact with each other through a process of institutional bricolage. Critical institutionalism and institutional bricolage are used to analyze the operationalization of fisheries co-management on Lake Victoria in East Africa, identifying how socially embedded institutions interact with the structures and processes of co-management, with implications for natural resource management and livelihoods.

Barnes in Chapter 6 takes a similar approach and also draws on critical institutionalism. She begins, however, with an introduction to commons scholarship, explaining how literature on mainstream institutionalism is increasingly recognizing non-deliberate, or socially embedded, institutions as inputs to purposeful, or bureaucratic, institutions. She argues, however, that this approach does not go far enough in terms of recognizing the influence of non-deliberate institutions and how they interact with purposeful institutions. Barnes analyzes a case study of community forest governance in Andhra Pradesh, India, to illustrate the role of non-deliberate institutions.

A further approach to analyzing natural resource governance is one that puts power, and the analysis of power, at its centre. This is political ecology, described in Chapter 8 by Upton as a loose body of theory, an approach or a frame of enquiry which recognizes that power relations influence people–environment relations. Upton applies a political ecology lens to a number of contemporary environmental issues, approaches and innovations in Mongolia, specifically in relation to conservation, carbon markets, and politics and mining, identifying how discourses underlying these challenge pastoralist identity and practices.

In Chapter 9, Aswani introduces the idea of there being multiple centres of decision-making, taking a polycentric lens to examine the multilevel and multi-centre system of customary sea tenure in the Western Solomon Islands. Through this approach, he identifies how institutions have evolved over time and how they explain a pluralistic and nested system of marine rights and governance. Aswani goes on to explain why and how understanding this diversity of rights and institutions matters for marine conservation efforts.

Chapter 10 focuses on the 'everyday practice' of natural resource governance, recognizing that governance in practice experiences a range of constraints and challenges that those involved have to navigate and cope with. Funder draws on literature on the anthropologies of the African state, as well as public administration studies, organization studies, development studies and anthropology. Through this, he investigates how everyday natural resource governance is carried out by the managers, technical specialists and extension agents who implement state policies, laws and projects on the ground, illustrating the insights gained from this approach through a case study of the implementation of climate change policies and projects by district government officers in Zambia.

Focus on the local is continued in Chapter 10, where Faye reviews conceptual tools for analyzing the legal and institutional effects of decentralization policies in natural resource management. From this, Faye investigates institutional 'choice and recognition' in decentralized forestry management in Senegal. He observes that state law is not a single legal domain but is characterized by multilayered institutions resulting from policies and projects which form parallel structures, rules and forms of accountability outside of the democratic system. The approach demonstrates how the plurality of legal systems, structures and institutions affects natural resource governance practice and outcomes.

Chapter 11 examines new modes of governance, focusing on PES – or Conservation Incentive Payment (CIP) – initiatives to offset the costs of living with carnivores and to make conservation beneficial to affected people. Pekor et al. analyze the interplay between governance and policy design, focusing in particular on the policy dimension of governance and its operationalization through policy instruments, drawing on the potential to implement a CIP program to foster human–lion coexistence in Ngorongoro, Tanzania. In addition to bringing attention to the interface between governance and policy instruments, the case study illustrates governance in practice through the co-design of a potential policy response to human–wildlife conflict.

The book closes with a concluding chapter (Chapter 12) in which the approaches covered in the book are compared in terms of the focus of analysis, methodologies and insights that emerge. Whilst different approaches were taken, focusing on different aspects of natural resource governance, a strong concern with institutions and local context emerged, reflected by a preference for qualitative research. The chapter clusters insights from the review of approaches and literature and analysis within three themes: multiplicity and diversity of institutions and multiplicity of influential variables; interactions between institutions; and governing in practice. The chapter concludes that diversity and flexibility, with inclusivity, are key for legitimate and effective governance systems and practice. This indicates that locally specific and adaptive institutions should be encouraged, with the remit, capacity and incentives for sustainable and equitable natural resource governance.

References

Adams, W. and Hulme, D. (2001) 'Conservation & community: Changing narratives, policies & practices in African conservation', in D. Hulme and M. Murphree (eds) *African Wildlife & Livelihoods. The Promise and Performance of Community Conservation*. James Currey, Oxford, pp. 9–23.

Adhikaria, B. and Di Falco, S. (2009) 'Social inequality, local leadership and collective action: An empirical study of forest commons', *European Journal of Development Research*, vol. 21, no. 2, pp. 179–194.

Agrawal, A. (2001) 'Common property institutions and sustainable governance of resources', *World Development*, vol. 29, no. 10, pp. 1649–1672.

Agrawal, A. (2003) 'Sustainable governance of common-pool resources: Context, methods, and politics', *Annual Review Anthropology*, vol. 32, pp. 243–262.

Agrawal, A. and Gibson, C. C. (1999) 'Enchantment and disenchantment: The role of community in natural resource conservation', *World Development*, vol. 2, no. 4, pp. 629–649.

Agrawal, A. and Ribot, J. (1999) 'Accountability in decentralization: A framework with South Asian and West African cases', *The Journal of Developing Areas*, vol. 33, no. 4, pp. 473–502.

Alonso-Población, E. and Siar, S. V. (2018) 'Women's participation and leadership in fisherfolk organizations and collection action in fisheries: A review of evidence on enablers, drivers and barriers', in *FAO Fisheries and Aquaculture Circular No. 1159*. Food and Agricultural Organization, Rome.

Anderson, J. and Mehta, S. (2013) *A Global Assessment of Community Based Natural Resource Management: Addressing the Critical Challenges of the Rural Sector*. USAID, Washington, DC.

Armitage, G. (2008) 'Governance and the commons in a multi-level world', *International Journal of the Commons*, vol. 2, no. 1, pp. 7–32.

Armitage, D., Berkes, F. and Doubleday, N. (eds) (2007) *Adaptive Co-management: Collaboration, Learning, and Multi-Level Governance*. UBC Press, Vancouver.

Batterbury, S. P. J. and Fernando, J. L. (2006) 'Rescaling governance and the impacts of political and environmental decentralization: An introduction', *World Development*, vol. 34, no. 11, pp. 1851–1863.

Baynes, J., Herbohn, J. and Dressler, W. (2016) 'Power relationships: Their effect on the governance of community forestry in the Philippines', *Land Use Policy*, vol. 54, pp. 169–176.

Benjaminsen, T. A. and Svarstad, H. (2010) 'The death of an elephant: Conservation discourses versus practices in Africa', *Forum for Development Studies*, vol. 37, no. 3, pp. 385–408.

Campese, J. (2016) 'Natural resource governance framework assessment guide: Learning for improved natural resource governance', *IUCN/CEESP NRGF Working Paper*. IUCN and CEESP, Gland.

Carlsson, L. and Sandström, A. (2008) 'Network governance of the commons', *International Journal of the Commons*, vol. 2, no. 1, pp. 33–54.

Cashore, B., Auld, G. and Newsom, D. (2004) *Governing through Markets: Forest Certification and the Emergence of Non-State Authority*. Yale University Press, New Haven, CT.

Chaffin, B. C., Gosnell, H. and Cosens, B. A. (2014) 'A decade of adaptive governance scholarship: Synthesis and future directions', *Ecology and Society*, vol. 19, no. 3, p. 56.

Cleaver, F. and Hamada, K. (2010) '"Good" water governance and gender equity: A troubled relationship', *Gender & Development*, vol. 18, no. 1, pp. 27–41.

Cornwall, A. (2008) 'Unpacking "participation": Models, meanings and practices', *Community Development Journal*, vol. 43, no. 3, pp. 269–283.

Cox, M., Arnold, G. and Tomás, V. S. (2010) 'A review of design principles for community-based natural resource management', *Ecology and Society*, vol. 15, no. 4, p. 38.

Ece, M., Murombedzi, J. and Ribot, J. (2017) 'Disempowering democracy: Local representation in community and carbon forestry in Africa', *Conservation and Society*, vol. 15, no. 4, pp. 357–370.

FAO. (2018a) *The State of World Fisheries and Aquaculture 2018—Meeting the Sustainable Development Goals*. FAO, Rome.

FAO. (2018b) *The State of the World's Forests 2018—Forest Pathways to Sustainable Development*. FAO, Rome.

Folke, C., Hahn, T., Olsson, P. and Norberg, J. (2005) 'Adaptive governance of social-ecological systems', *Annual Review of Environmental Resources*, vol. 30, pp. 441–473.

Franks, T. and Cleaver, F. (2007) 'Water governance and poverty: A framework for analysis', *Progress in Development Studies*, vol. 7, no. 4, pp. 291–306.

Galvin, K. A., Beeton, T. A. and Luizza, M. W. (2018) 'African community-based conservation: A systematic review of social and ecological outcomes', *Ecology and Society*, vol. 23, no. 3: 39. doi.org/10.5751/ES-10217-230339.

Giessen, L., Burns, S., Sahide, M. A. K. and Wibowo, A. (2016) 'From governance to government: The strengthened role of state bureaucracies in forest and agricultural certification', *Policy and Society*, vol. 35, pp. 71–89.

Gisselquist, R. M. (2012) 'Good governance as a concept, and why this matters for development policy', *Working Paper No. 2012/30*. UNU-WIDER, Helsinki.

Grindle, M. (2004) 'Good enough governance: Poverty reduction and reform in developing countries', *Governance: An International Journal of Policy, Administration and Institutions*, vol. 17, no. 4, pp. 525–548.

Grindle, M. (2007) 'Good enough governance revisited', *Development Policy Review*, vol. 25, no. 5, pp. 553–574.

Hyden, G., Court, J. and Mease, K. (2004) *Making Sense of Governance: Empirical Evidence from Sixteen Developing Countries*. Lynne Reiner, London.

Jones, B. (2010) 'Ostrom and Namibian conservancies', *Current Conservation*, vol. 4, no. 3, pp. 21–23.

Karnad, D. (2017) 'Navigating customary law and state fishing legislation to create effective fisheries governance in India', *Marine Policy*, vol. 86, pp. 241–246.
Kooiman, J. (2003) *Governing as Governance.* Sage, London.
Kooiman, J., Bavinck, M., Chuenpagdee, R., Mahon, R. and Pullin, R. (2008) 'Interactive governance and governability: An introduction', *The Journal of Transdisciplinary Environmental Studies*, vol. 7, no. 1, pp. 1–11.
Leisher, C., Temsah, G., Booker, F., Day, M., Samberg, L., Prosnitz, D., Agarwal, B., Matthews, E., Roe, D., Russell, D., Sunderland, T. and Wilkie, D. (2016) 'Does the gender composition of forest and fishery management groups affect resource governance and conservation outcomes? A systematic map', *Environmental Evidence*, vol. 5, no. 1, p. 6.
Lockwood, M., Davidson, J., Curtis, A., Stratford, E. and Griffith, R. (2010) 'Governance principles for natural resource management', *Society and Natural Resources*, vol. 23, no. 10, pp. 986–1001.
Lund, J. F. and Saito-Jensen, M. (2013) 'Revisiting the issue of elite capture of participatory initiatives', *World Development*, vol. 46, pp. 104–112.
Matzdorf, B., Sattler, C. and Engel, S. (2013) 'Institutional guides and governance structures of PES schemes', *Forest Policy and Economics*, vol. 37, pp. 57–64.
Muyengwa, S., Child, B. and Lubilo, R. (2014) 'Elite capture: A comparative case study of meso-level governance in four southern Africa countries', in G. Barnes and B. Child (eds) *Adaptive Cross-Scalar Governance of Natural Resources.* Routledge, London.
Mwangi, E. and Wardell, A. (2012) 'Multi-level governance of forest resources', *International Journal of the Commons*, vol. 6, no. 2, pp. 79–103.
North, D. C. (1990) *Institutions, Institutional Change and Economic Performance.* Cambridge University Press, Cambridge.
Nunan, F. (2015) *Understanding Poverty and the Environment: Analytical Approaches and Frameworks.* Routledge, London.
Nunan, F. (2018) 'Navigating multi-level natural resource governance: An analytical guide', *Natural Resources Forum*, vol. 42, no. 3, pp. 159–171.
Nunan, F., Menton, M., McDermott, C. and Schreckenberg, K. (2018) 'Governing for ecosystem health and human wellbeing', in K. Schreckenberg, G. Mace and M. Poudyal (eds) *Ecosystem Services for Poverty Alleviation: Trade-Offs and Governance.* Routledge, London, pp. 159–173.
Ojha, H. R., Ford, R., Keenan, R. J., Race, D., Vega, D. C., Baral, H. and Sapkota, P. (2016) 'Delocalizing communities: Changing forms of community engagement in natural resources governance', *World Development*, vol. 87, pp. 274–290.
Ostrom, E. (1990) *Governing the Commons.* Cambridge University Press, New York.
Ostrom, E. (2007) 'A diagnostic approach for going beyond panaceas', *Proceedings of the National Academy of Sciences*, vol. 104, pp. 15181–15187.
Ostrom, E., Dietz, T., Dolsak, N., Stern, P. C., Stonich, S. and Weber, E. U. (eds) (2002) *The Drama of the Commons.* National Academies Press, Washington, DC.
Pahl-Wostl, C. (2009) 'A conceptual guide for analysing adaptive capacity and multi-level learning processes in resource governance regimes', *Global Environmental Change*, vol. 19, pp. 354–365.
Persha, L. and Andersson, K. (2014) 'Elite capture risk and mitigation in decentralized forest governance regimes', *Global Environmental Change*, vol. 24, pp. 265–276.
Poteete, A. (2012) 'Levels, scales, linkages, and other "multiples" affecting natural resources', *International Journal of the Commons*, vol. 6, no. 2, pp. 134–150.
Poteete, A. R. and Ribot, J. C. (2011) 'Repertoires of domination: Decentralization as process in Botswana and Senegal', *World Development*, vol. 39, no. 3, pp. 439–449.

Ribot, J. C., Lund, J. F. and Treue, T. (2010) 'Democratic decentralization in sub-Saharan Africa: Its contribution to forest management, livelihoods, and enfranchisement', *Environmental Conservation*, vol. 37, no. 1, pp. 35–44.

Secretariat of the Convention on Biological Diversity (2014) *Global Biodiversity Outlook 4*. Secretariat of the Convention on Biological Diversity, Montréal.

Shackleton, C. M., Willis, T. J., Brown, K. and Polunin, N. V. C. (2010) 'Reflecting on the next generation of models for community-based natural resources management', *Environmental Conservation*, vol. 37, no. 1, pp. 1–4.

Springer, J. (2016) *Initial Design Document for a Natural Resource Governance Guide*, IUCN/CEESP NRGF Working Paper. IUCN and CEESP, Gland.

Tadie, D. and Fischer, A. (2017) 'Natural resource governance in lower Omo, Ethiopia – negotiation processes instead of property rights and rules?', *International Journal of the Commons*, vol. 11, no. 1, pp. 445–463.

Termeer, C. J. A. M., Dewulf, A. and van Lieshout, M. (2010) 'Disentangling scale approaches in governance research: Comparing monocentric, multilevel, and adaptive governance', *Ecology and Society*, vol. 15, no. 4, p. 29.

Williams, B. K. (2011) 'Adaptive management of natural resources – Framework and issues', *Journal of Environmental Management*, vol. 92, pp. 11346–11353.

World Bank (1989) *Sub-Saharan Africa: From Crisis to Sustainable Growth*. World Bank, Washington, DC.

2

MANAGEMENT IN THE GUISE OF GOVERNANCE? RETHINKING THE ENDS AND THE MEANS OF NATURAL RESOURCE GOVERNANCE

Prakash Kashwan

Introduction

Theories and discourses of governance have undergone a fundamental transformation in last few decades. Multilateral agencies, such as the World Bank and international development agencies, such as the Department for International Development (DFID), have argued that the essence of governance must shift from top-down regulations toward fostering relationships of interdependence among various stakeholders. They argue that the focus must shift from governments or government agencies to a plurality of actors and agencies in the state and society that contribute, in a variety of ways, to the processes and outcomes of governance. The World Bank's Governance Practice is 'focused on the critical interface between the public and private sectors, recognizing public sector risks that may prevent investors from entering a country' (World Bank, 2019).

A rapid diffusion and adoption of a framework of 'governing without government' (Rhodes, 1996), or with a minimal role for the state, has led to an industry of frameworks and indices to measure the success of governance in the form of long lists of disparate indicators. The World Bank's Worldwide Governance Indicators (WGI) comprises six aggregate indicators: (1) voice and accountability; (2) political stability and the absence of violence; (3) government effectiveness; (4) regulatory quality; (5) the rule of law; and (6) control of corruption (World Bank, 2007b: 2 cited in Gisselquist, 2012: 4). Lockwood et al. (2010: 997) developed a list of eight principles of natural resource governance (legitimacy, transparency, accountability, inclusiveness, fairness, integration, capability, and adaptability), similar to the World Bank's governance indicators.

Despite the expansive list of indicators included in various governance indices, they do not help answer some of the most fundamental questions as to governance for what, by whom and under what conditions. Similarly, the indicators-dominated governance scholarship says little about the relational aspects of governance (e.g., who decides what the goals of governance in a particular context are, who designs the mechanisms and systems of governance in pursuit of specific goals, and who monitors whether such pursuits are implemented). The omission of these questions is surprising because they are as old as the scholarly and policy literatures on governance. In his pioneering article entitled 'The New Governance: Governing without Government,' Rhodes (1996: 662) argued that the process of hollowing-out of the state creates an 'institutional complexity' that 'obscures who is accountable to whom for what.' These questions also come up time and again in the applied literature on governance in specific sectors (e.g., health, education, fisheries, and forests) (Nunan, 2006).[1] Similar questions are also asked in the context of other specific aspects of governance (e.g., legitimacy and representation) and occasionally vis-à-vis the issues of 'regulatory quality' and 'rule of law.' In most cases, though, such analyses do not feed back into the conceptual apparatus of governance, which seems to be stuck in the multi-criteria indices that a prominent governance expert has described as 'a set of well-meaning but problematic proverbs' that are devoid of serious theoretical foundations (Andrews, 2008: 379).

Instead of producing another list of governance principles or requirements, in this chapter I propose a meta-framework comprising five fundamental aspects of governance, which I argue must inform the development of architectures related to the governance of specific sectors. My goal here is to create a theoretically grounded, yet accessible, diagnostic framework that shines light on the most salient features of a governance regime. A framework enables a researcher to identify key variables, factors, processes and relationships relevant to a specific inquiry (Schlager, 2007). It helps organize inquiries for studying processes and explaining outcomes at one or more of the micro, meso, or macro levels without being constrained by one specific theory. Frameworks can be used for diagnostics purposes or for making predictions and policy prescriptions (Ostrom, 2007). For example, the institutional analysis and development (IAD) framework, which is among the most frequently used frameworks for institutional analysis in the fields of environment and development, can be used for both diagnostic and prescriptive purposes.[2]

In the following section, I offer a selective overview of the governance literature to justify the development of the meta-framework. The meta-framework proposed here is based on a theoretically motivated reading of the governance scholarship. Moreover, this framework was developed without prejudice to the contents of the three natural resource governance proposals that are the subject of the analysis presented in these pages. As such, the meta-framework proposed here responds to important theoretical considerations, without being constrained by the boundaries set by the applied research on natural resource governance. This is followed by an analysis of the extent to which individual elements of governance included in the meta-framework proposed here are accounted for in each of the following three distinct natural resource governance frameworks:

(1) Resource Governance Index (RGI) developed by Natural Resource Governance Institute (NRGI), (2) Natural Resource Governance Framework (NRGF) developed by the International Union for Conservation of Nature (IUCN), and (3) the memorandum on 'Resource Politics for a Fair Future' (RPFF) developed by the Heinrich Böll Foundation (HBF). These three approaches have been selected because they meet a number of analytically helpful criteria. One, taken together, these three approaches cover a broad range of natural resource policy sectors. Second, they have been developed and promoted by well-respected think tanks and advocacy groups with significant influence over global debates on natural resource governance. Third, and perhaps most importantly, each of these frameworks has been designed from the perspective of actors with varied goals and commitment. Such diverse motivations for resource governance brings to light the competing normative concerns and goals that often inform natural resource governance.

Insights gained from an application of the meta-framework to an analysis of the three selected natural resource governance proposals are then presented. In the concluding section, I reflect on the lessons learned from this analysis and recommendations for future research on natural resource governance.

A framework for the assessment of governance architecture

The arguments and analyses of key aspects of governance that constitute the meta-framework proposed here are based on a political economy of institutions (PEI) perspective that I have developed elsewhere (Kashwan, 2017). The PEI takes as its point of departure a critical understanding of institutions and governance arrangements, which suggests that policies, institutions, and organizational arrangements in a governance regime reflect the balance of power among different constituents in a society (Knight, 1992; Moe, 2005). While recognizing the value of appropriately designed policies, programs, and institutions, a political economy approach does not take for granted the emergence of an 'objective' or a 'neutral' governance architecture that is expected to be universally good. Instead, the PEI framework posits that the emergence of specific governance regimes and their translation into the outcomes of resource governance are contingent on whether the actors and agencies with significant political power have proper incentives to support such processes and outcomes. Instead of assuming a reified model of individuals expected to behave as *homo economicus*, the meta-framework I propose recognizes the embeddedness of individuals within specific political and economic contexts and outlines a system of checks and balances that would enable individuals and groups of individuals to overcome the constraints that such embeddedness imposes on them. Actors and agencies contribute to desired governance outcomes when they are able to overcome the expectations attached to a robot model of *homo economicus* or that of an embedded helplessness. More broadly, the proposed framework offers a hybrid model of governance that combines elements of hierarchical and horizontal governance models.

I use Merilee Grindle's definition of governance as 'the institutions, systems, 'rules of the game' and other factors that determine how political and economic interactions are structured and how decisions are made and resources allocated' (Grindle, 2010: 2). This definition does not presume or imply specific normative goals (e.g., transparency, legitimacy, and inclusion) which are desirable but cannot be accomplished without putting in place a suitable governance architecture. The meta-framework helps redirect the scholar's analytical gaze on such features of governance. On the other hand, Grindle's and several other commonly used definitions of governance often omit an important dimension, which is that governance is about steering by setting the course, putting in place mechanisms for monitoring progress and, in most cases, occasional course correction (Crawford, 2006: 453). The meta-framework proposed here fills this gap by recognizing the importance of stating upfront the goals of specific governance arrangements, while fully recognizing the legitimacy of a diversity of goals pursued by different actors, agencies and governance arrangements. To clarify, 'steering' should not be equated to top-down imposition of the means and ends of governance. Instead, it entails creating the right kind of conditions that would allow various stakeholders to address the problems of free-riding, while developing a trust in the viability of accomplishing governance that enjoys broad-based support among the key stakeholders.

The governance meta-framework I develop here comprises five key dimensions intended to inform governance assessments: (1) A clear specification of the goal/s of governance; (2) Representation of different stakeholders; (3) Monitoring and enforcement of socially agreed on sets of rights and responsibilities (an aspect of governance that is especially salient in the area of natural resource governance); (4) The commitment of state power as the ultimate guarantor of institutional integrity; and (5) Institutionalization of powers and counterpowers for accountability in governance. In the following, I develop the rationale for and the main advantages of the selection of the five key dimensions of the meta-framework.

Specification of goals

Governance strategies and architecture cannot be conceptualized in isolation from the specific goals that a governance regime is meant to pursue. For instance, in fairly competitive market settings, the promotion of efficient industrial production can be accomplished through market-based arrangements. However, if the goal is to promote environmental protection, a laissez-faire framework of governance is unlikely to be effective. This is so because the production of environmental externalities is one of the main examples of market failure. Accordingly, neither the suitability nor the effectiveness of a specific governance regime can be ascertained without reference to specific goal/s, or a sequence of prioritization among varied and competing goals of governance.

A clear definition and statement upfront of the goals of governance serves a number of important ends. One, as mentioned above, making transparent the goal of governance allows for proper deliberations and analyses about whether a set of governance strategies under consideration are likely to serve the stated goals (e.g., resource conservation, an economically efficient or socially just management of natural resources). Two, a discussion of a plurality of governance goals also points to the heterogeneity of preferences and interests among various stakeholders, which in turn brings to fore the challenge of developing multi-stakeholder governance arrangements. For example, stakeholders interested in the varied goals of promoting biodiversity and wildlife conservation, carbon credits, and local livelihoods could be brought together on a common platform, but it would be nearly impossible to simultaneously promote each of these goals with equal efficacy. Moreover, in many cases, the stated prioritization of goals may not translate into such prioritization in practice because of the power and influence that certain actors are able to wield in promoting their ideas and interests. Under some circumstances, some of those who do not share the stated goals of resource governance may seek to undermine the success of governance strategies lest the success of such governance arrangements undermines their interests. Accordingly, the meta-framework we propose here includes transparent specification of the goals of governance as an important prerequisite for a proper assessment of the efficacy of natural resources governance with a specific configuration of actors, agencies and institutions.

Representation of key stakeholders

The popularity of the concept of networked or polycentric governance is owed to the assumption that the presence and participation of non-state actors facilitates a 'general representation' of the varied interests in a society (Boltanski and Chiapello, 2005, cited in Davies, 2011: 1). Representation of a plurality of interests makes it more likely that information relevant to the determination of governance functions is readily available. This is especially so in cases in which a reasonable distribution of political and economic resources allows actors with varied interests to act as checks against the dominance of a narrow set of actors or agencies. In those cases, the pursuit of multiple and competing interests may foster some level of accountability of key actors and agencies charged with overseeing the success of governance arrangements. However, such a well-balanced representation of varied interests cannot be assumed a priori in either hierarchical or horizontal governance models. Specific institutional arrangements – pertaining to both vertical and horizontal relations – are needed to foster effective representation.

Considering the high levels of political and economic inequalities, in many cases, hierarchical arrangements may be needed for ensuring an effective representation of economically weaker and politically marginalized groups (Bardhan, 2002). A consideration of stakeholder representation as an important dimension of governance points to concrete and observable empirical implications that can be recorded and analyzed comparatively in a variety of settings

(Kashwan, 2017). An empirical analysis of the representation of varied interests also offers a window into the distributional consequences of a governance regime, which is important for instrumental as well as normative considerations of equity and social justice.

Institutional arrangements for monitoring and enforcement

Natural resources are subject to the grave risks of negative externalities, that is, situations in which producers or consumers impose costs that are not accounted for in pricing mechanisms.[3] For instance, multinational corporations can sell electronic and computer products much cheaper than they could if they were required to pay for the proper disposal of toxic waste generated in the process of manufacturing such goods, say, in parts of China.[4] By letting their toxic wastes accumulate in open-air 'toxic lakes,' the manufacturers benefit their customers, while imposing negative externalities on local residents, a majority of who are from poor and marginalized social groups. Similarly, the poorly regulated greenhouse gas emissions from a variety of outsourced manufacturing operations in China and elsewhere impose externalities on humanity at large.

Economists often refer to 'externalities' as unintended consequences. However, from an institutional perspective, negative externalities are a product of perverse incentives that motivate producers to impose environmental costs on the society at large (see Vatn and Bromley, 1997). From a societal perspective, externalities are neither unavoidable nor unintended. Unsustainable extraction of natural resources is a major risk in any part of a supply chain, irrespective of whether the supply chain is controlled by the state or market actors. To address these risks, any natural resource governance architecture must include robust institutional arrangements for monitoring and enforcement of regulations intended to curb externalities and prevent environmental degradation. While the responsibility for monitoring specific activities in specific areas can be outsourced to non-state actors, as discussed below, the state has a norm-setting role of a sheet anchor.

Credible commitment of state power

'Rule of law' is the basic building block for all types of institutional arrangements, the success of which depends almost entirely on the affected actors being assured of a jurisdiction-wide enforcement of rules that a society has collectively agreed on. In the absence of credible mechanisms for 'graduated sanctions' (Ostrom, 1990) against behaviors that undermine the collectively agreed on goals of governance, some actors or agencies are likely to be inclined to free-ride. The extent of free-riding is likely to be directly proportional to the extent to which potential defaulters believe that there would not be any repercussions. Once such perverse incentives have been set in motion, individuals and groups of individuals, including those interested in abiding by the norms of governance, would find it difficult to do so. This draws our attention to the crucial role of the rule of law, broadly understood, in maintaining the essential integrity of a rule-based framework of governance.

Rule of law requires the support and cooperation of a variety of state and non-state actors, including citizen groups who may be involved in the local provisions of conflict resolution mechanisms. As such, the state is not the only source of legitimacy for a rule of law framework and its enforcement. Yet only the state enjoys a legitimized monopoly over coercive actions in a rule of law framework. Accordingly, the state's commitment is indispensable for ensuring the credibility of the rule of law, especially in the face of market failures and other externalities. Potential offenders would be emboldened greatly in the face of the weakness of the state's commitment to a rule-of-law framework and its impartial enforcement because, to wit, the buck stops with the state. This simple insight is often neglected in the governance scholarship originating in the global North because the essential integrity of the state's commitment to enforcement of a rule of law is taken for granted there. However, the enforcement of the rule of law cannot be taken for granted in many settings, including in the global North when it comes to corporate culpability in the areas of environmental degradation and pollution.[5]

The barriers against the integrity of rule of law are especially serious in some of the resource-rich economies that are characterized by poor institutional infrastructure, a weak rule of law and problems of corruption at the highest levels of political hierarchy (Kolstad and Søreide, 2009). As a result, an effective governance framework must include explicit arrangements to ensure the state's very visible commitment to the enforcement of socially legitimate governance arrangements. The focus on social legitimacy precludes support for evidently exploitative governance arrangements (e.g., the cases in which valuable natural resources are squandered away by state and non-state actors engaged in rent-seeking practices or in cases where the colonial legacies of extractive resource management enable a concentration of authority in the hands of powerful state agencies that are not accountable to the society at large). Not only does a lack of accountability undermine the legitimacy of natural resource governance, but it is also likely to undermine the sustainability of natural resource systems.

Institutionalization of powers and counterpowers

'Power' is central to the functioning of any governance arrangements. The World Bank defines governance as 'the exercise of political power to manage a nation's affairs' (World Bank, 1989: 61). In another publication that is cited frequently in the scholarship on governance, the Bank defines governance as 'the manner in which power is exercised in the management of a country's economic and social resources' (World Bank, 1994: xiv). Yet none of the major World Bank publications accounts for how power asymmetries may undermine the optimism that the Bank and other international agencies attach to the agenda of good governance.[6] The significance of this omission is acknowledged by Gerry Stoker, a prominent governance scholar, who argues that the pioneering governance scholarship by Rhodes and Stoker failed to account for the uneven distributions of 'hard' and 'soft' power among the main participants in networked governance (Stoker, 2011: 28).[7] Such neglect of the questions of power within the literature on governance is owed, in part, to the assumption among many scholars that networked governance produces 'dispersion and disorganization of power and institutions' (Boltanski and Chiapello, 2005: 138; cited in Davies, 2011: 2). These assumptions are clearly invalid in a large number of cases. Accordingly, in this era of increasing inequalities, it is important to ask whether a system of governance institutionalizes the distribution of powers and counterpowers in the hands of a plurality of state and non-state actors (for the development of this argument in the context of decentralization, see Agrawal and Ribot, 1999). Moreover, such a system of distributed power would work only if each of the power-holders is capable of censuring actions that undermine the integrity of governance arrangements, especially when the violator happens to be a state agency or an actor with authoritative powers. A failure on this front raises the age-old question, 'Who will guard the guardians?' (Hurwicz, 2008).

To sum up, a parsimonious meta-framework that I propose here for the purpose of assessing the robustness of a governance system comprises the following five key elements that draw our attention to a transparent specification of governance goals, the representation of key stakeholders, arrangements for monitoring and enforcement, credible commitment of state power and institutionalization of powers and counterpowers to prevent abuses at the hands of powerful actors in the state, society and markets. While the discussion of each of the five elements has been presented separately for the sake of analytical clarity, each of these elements also supports others. Taken together, they form an ecosystem of actors, institutions and agencies whose behaviors and actions contribute to the pursuit of specific goals that a governance system is meant to accomplish.

Applying the meta-framework to three natural resource governance frameworks

I apply the meta-framework developed above to critically engage with and analyze three prominent frameworks for natural resource governance: (1) RGI developed by NRGI; (2) NRGF developed by IUCN; and (3) RPFF, a memorandum of the HBF (henceforth, the memo).

RGI developed by NRGI

RGI comprises three main components: (i) revenue management; (ii) value realization; and (iii) enabling environment. The first two indicators produce sector-specific measures of governance at the national level, while the third component reflects indicators that Michener (2015) would define as concept indicators. The component 'enabling environment' includes measures of transparency, regulatory quality, rule of law, and government effectiveness that are based on indicators of governance more generally at the national level. That is, the data on this third set of indicators relates to a country as a whole, without reference to a specific resource governance sector, with a necessary underlying assumption that if a country scores well on governance in general, it must also do well on the question of resource governance. To evaluate RGI on the basis of the meta-framework developed here, I examine how each of the three components of index is measured in practice.

The specification of goals

RGI defines resource governance as 'the rules, disclosures, oversight procedures and enabling environment that allow citizens to hold their government to account *for managing their extractive resource wealth*' (NRGI, 2017a: 1; italics added for emphasis). The RGI has already narrowed, even if implicitly, the goals of natural resource governance as the management of natural resources primarily for the purpose of extraction and the management of the wealth earned from resource exploitation. This is a significant departure from the principles espoused in RGI's founding document, the Natural Resource Governance Charter, which states that resource governance could be designed to pursue a variety of social, cultural, and economic goals.

The narrowing of governance goals in RGI implicitly confounds the values of transparency in governance. More importantly, by privileging one specific goal over equally legitimate goals of social equity and environmental protection, the RGI also undermines the goals of representation of diverse interests in resource governance.

The representation of a diversity of interests

The closest the RGI comes to account for representation of varied interests is in its discussion of 'civic space,' which appears alongside transparency in the RGI's accounting of 'citizens seeking to hold their governments to account' (NRGI, 2017b: 21). The role of civic space is envisaged mainly with reference to the transparency component of RGI, such that it focuses on 'citizens' freedom to use disclosures to hold their governments accountable' (ibid.). This is symptomatic of RGI's reliance on legal and informational environment, without accounting for who makes crucial governance decisions to what ends. This is also in sync with a second definition of natural resource governance that the RGI documents cite, in which governance is 'the institutions, rules and practices that determine *how company executives and government officials make decisions* and engage and affect citizens, communities and the environments they inhabit' (NRGI, 2017b: 3, italics added for emphasis). Notice that corporations are part of the decision-making process along with government officials, who 'engage and affect' citizens.

A second way in which the RGI touches on questions of representation is in accounting for whether national laws require that resource revenues are shared with subnational governments. Subnational revenue sharing could be considered one way of accounting for the interests of local residents. However, the subcomponent of 'subnational resource revenue sharing' is 'only assessed in countries that have these institutions' (NRGI, 2017a: 4). In other words, if a country did not have provisions for subnational revenue sharing, it would not affect its index scores.[8] Moreover, even in countries that do have revenue sharing laws and stipulations, the RGI does not account for whether such laws translate into material benefits for citizens at large. Scholarship on natural resource decentralization shows us that citizens are unlikely to benefit much if national governments have not devolved meaningful and discretionary authority in the hands of subnational governments (Agrawal and Ribot, 1999; Ribot et al., 2006). Accordingly, the RGI accounts for representation of the interests of diverse social constituencies only marginally and rather unevenly across the sample of countries for which the data on RGI is available.

Monitoring and enforcement

The RGI includes 'regulatory quality' as one of the indicators of the index component 'Enabling Environment.' The subcomponents that add up to the RGI score on regulatory quality are borrowed from the World Bank's WGI. The WGI define 'regulatory quality' as an aggregate indicator of 'perceptions of the ability of the government to formulate and implement sound policies and *regulations that permit and promote private sector development*' (italics added for emphasis).[9] From a private sector perspective, regulatory quality is about minimal regulations that help promote private exploitation of natural resources. On the other hand, from an environmental perspective, regulations would entail institutionalizing checks against private incentives potentially linked to environmental harms and environmental externalities. To assess how these differing perspectives influence RGI measurements, it is useful to consider two subcomponents of the index pertaining to 'local environmental impacts' and 'subnational resource revenue sharing' (NRGI, 2017b: 15–16).[10]

Measurement of local impacts draws on answers to questions such as, 'Are there *rules governing compensation to landowners*?' and 'Are there specific *procedures governing the resettlement of land users*?' Environmental mitigation aspects of local impact are captured in the question whether '*environmental mitigation management plans been publicly disclosed*.' As such, the component on 'value realization,' which 'measures the quality of governance around exploration, production, environmental protection, and revenue collection' should probably be labeled as the *promise* of value realization. Similarly, each of the questions included within the second major index component, 'revenue management,' is based on an assessment of the existence of laws and rules governing specific aspects of governance or formal statements of information disclosure. In fact, a survey of the RGI questionnaire used for the 2017 RGI report suggests that barring a few indicators, such as whether government's claims about subnational revenue transfers were externally audited, nearly all questions that the RGI labels as 'Practice Questions' are also questions about disclosure of relevant information.[11] Most of the RGI index components thus do not measure the 'practice' of rule enforcement, that is, whether and to what extent specific rules are followed in practice.

Commitment of state power

As discussed above, the main thrust of RGI is toward facilitating private sector investment in the task of resource extraction and proper management of resource wealth at the national level. RGI's main components and subcomponents are focused on ensuring credible state commitments to the corporations engaged in natural resource-based industries. The RGI focused mainly on one of the twelve percepts of the Resource Governance Charter, which suggests that 'the (national) government should encourage efficient exploration and production operations, and allocate rights transparently' (NRGI, 2014: 13). As such, the RGI rewards states that afford sufficient protections to businesses and corporations, without being bound by the strong regulatory role of the state focused on social and environmental goals.

The RGI's emphasis on the role of the state as 'market-protecting state' is quite different from the role of the state conceptualized in the meta-framework, which is that of acting as the sheet anchor for the rule of law and as a trustee of the social and environmental interests within the extractive sector. This is not just an academic pipe dream, as evident from a reading of the Natural Resource Charter, which also includes the following precepts:

> Precept 5: The government should pursue opportunities for local benefits, and account for, mitigate and offset the environmental and social costs of resource extraction projects.
>
> Precept 7: The government should invest revenues to achieve optimal and equitable outcomes, for current and future generations.
>
> Precept 9: The government should use revenues as an opportunity to increase the efficiency of public spending at the national and sub-national levels.
>
> Precept 11: Companies should commit to the highest environmental, social and human rights standards, and to sustainable development.

None of these precepts are translated into substantive indicators within the RGI. The RGI is especially less attentive to the questions of environmental impact, as evident from the scoring scheme that takes 'some costs to society' as granted even for the top performing category of 'good' governance (a score $\geq$ 75). None of the categories specifically mention environmental impacts, the concerns that are set aside in the favor of the core focus on whether 'resource extraction can help society' (NRGI, 2017b: 5). As discussed previously, a system of natural resource governance is unlikely to ensure environmental protections unless the state and its machinery is committed to the enactment and enforcement of regulatory mechanisms meant to address market failures in the realm of social and environmental protections, including negative externalities of various types.

Powers and counterpowers

The RGI borrows the system-wide governance indicators from the WGI, which are based on the expert judgments about the overall state of affairs in a country's governance regime. These indicators are not specific to the resource extraction sectors (e.g., the mining and oil sectors) that the other components of RGI are focused on. As such, RGI does not measure whether government ministries, agencies, and officials are held to account on the specific questions related to resource governance. As discussed above, it also does not account for whether government agencies and officials monitor performance and the enforcement of sanctions against the erring license holders, including national and international corporations. The RGI does not consider also how the balance of power or lack thereof among key stakeholders affects the accountability of state and non-state actors. Instead, the centralization of authority in the hands of governments to decide on resource extraction and manage resource revenues, which is implied in the RGI, is likely to produce varied implications for different national and international actors. While it may be convenient for a multinational company to deal with one centralized authority, all else equal, the centralization of power and control is likely to undermine citizens' efforts to secure accountability of government agencies or state-owned corporations.

NRGF developed by IUCN

The NRGF is designed to promote effective and equitable natural resource governance – '*emphasizing rights-based approaches, equity and social justice*' (Springer, 2016: 5, italics added for emphasis). Such a broad and ambitious normative agenda leads the NRGF authors to propose an extensive list of principles including,

> Inclusive decision-making, Recognition and respect for tenure rights, Embracing diverse cultures and knowledge Systems, Devolution, strategic vision and direction, Empowerment, Coordination and Coherence, Sustainable Resources and Livelihoods, Social and Environmental Accountability, Special attention to the Vulnerable, Rule of Law, and Access to Justice.
>
> *(Springer, 2016: 7)*

The similarity of several of these principles to the components used in WGI and the RGI discussed above is noticeable. However, NRGF's definitions of key concepts is more broad-based. For instance, its definition of the rule of law accounts for the questions of equity and justice. Moreover, instead of defining specific criteria and indicators, the NRGF lists three macro categories: (i) legal/policy frameworks; (ii) implementation: institutions and processes; and (iii) outcomes, which seek to measure the extent to which governance principles and criteria are realized in practice (Springer, 2016: 10).

The framework document is accompanied by an extensive framework assessment guide (Campese et al., 2016), which further recognizes the importance of different contexts/concerns (conservation of particular resources or ecosystems, practice of extractive industry or large scale development, etc.), governance at multiple levels (local to national and regional/transboundary), with multiple objectives (learning, internal strengthening, advocacy, accountability), and by different actors (communities; private actors such as nongovernmental organizations (NGOs), individual landowners, or businesses; government agencies; or various combinations of any of these). The assessment guide lays out several dozen options and decisions that anyone planning for a governance assessment would have to make before they undertake an assessment.

Specification of goals

NRGF recognizes the potential for multiple perspectives on governance, depending on who leads an assessment (which is one of the meta-questions included within the NRGF assessment guide). While these considerations are supposed to inform an NRGF-inspired assessment, an evaluation of the clarity of goals is not part of the assessment itself. This may be attributed in part to the fact that NRGF is designed with specific goals; it is meant to assist in 'promoting and supporting a *rights based approach* to *conservation* in IUCN' (Springer, 2016: 5, italics added for emphasis). Noticeably, the rights-based approach is used as an instrument of conservation, which seems to be the ultimate goal of natural resource governance for IUCN.[12] NRGF is meant to 'strengthen the attention to governance dimensions in assessment processes focused on conservation or environmental objectives' (Springer, 2016: 12). True to its mandate, the IUCN remains focused on conservation, though it seeks to address, in its own ways, the question of equity in conservation.

Representation of stakeholders

IUCN offers one of the most ambitious definitions of governance, which includes a consideration of 'how power and responsibilities over natural resources are exercised, how decisions are taken, and how citizens – women, men, indigenous peoples and local communities – participate in and benefit from the management of natural resources.'[13] This sense of limiting citizen roles is also reinforced in how 'devolution' – the political powers for discretionary decision-making – which is one of the core NRGF principles, is qualified in the framework's founding text. Placed fourth in the list of the proposed NRGF principles, devolution appears with a hyphenated qualifier: 'Devolution – *especially for community-based natural resource governance*.' This means that devolution is not a focus, or at least explicitly so, in the management of nature protected areas, which constitute an overwhelming majority of IUCN interventions. Importantly, most of the world's protected areas continue to be governed not just in a fortress conservation mode (Brockington, 2002), or in a neoliberalized format (Dressler and Roth, 2011), but via a new ultra-aggressive approach often referred to as 'militarized conservation' (Duffy, 2016). NRGF does not recognize these major challenges, even as it offers a qualified inclusion of devolution applicable to a very small number of global conservation interventions.

Monitoring and enforcement

The NRGF is relatively less attentive to the specific tasks of monitoring and enforcement, which are captured under the generic rubric of the 'rule of law.' The NRGF assessment guide offers more details about the sources of relevant information, which include 'reports on the implementation and enforcement of laws, policies, rules, agreements, such as local records of natural resource-related fines, reports on Multilateral Environmental Agreements (MEAs) and human rights instruments, project monitoring reports and any related grievances, court cases, etc.' (Campese et al., 2016: 16). These sources of information are meant to assist those deploying NRGF for an assessment of natural resource governance. NRGF macro components include the laws and statutes as important anchors of governance but do not explicitly recognize the importance of monitoring and enforcement in natural resource governance. This is somewhat unexpected, since IUCN's main mandate is to promote conservation of resources, and, as discussed above, it overrides the goals of equity.

Commitment of state power

Out of the five key components of the meta-framework proposed here, the role of the state is the most neglected aspect of NRGF. A consideration of the state's role is implicit in the NRGF components related to accountability, the rule of law, protection of rights, and access to justice. However, NRGF omits a systematic discussion of the state's role and the ways in which it can be utilized to promote sustainable and equitable management of resources. This is not a minor omission because the state exercises legal ownership and control over nearly all PAs, except for a few that are situated either in legally recognized indigenous territories or even a smaller number of PAs that are formally recognized as community-conserved areas (CCA). Such overwhelming dominance of the state is also the source of problems of corruption and lack of accountability in the areas of nature conservation and natural resource governance more broadly (Kashwan, 2013). Such a status quo, full of various types of abuses of power,[14] makes it challenging to recognize the important theoretical role that the state and its authority plays in creating the foundations for the rule of law, which NRGF recognizes as one of its core principles.

Powers and counterpowers

The founding text of the NRGF recommends the building of 'structures and capacities' needed for people to 'hold governments, the private sector and other actors with roles and authority for natural resource governance responsible for their actions' (Springer, 2016: 9). Additionally, the document also explicitly states the need for 'a horizontal process in which *power dynamics are re-balanced*' (ibid. 8, italics added for emphasis), which as explained in Section 1, is an important prerequisite for the success of governance arrangements. Even so, the NRGF does not offer tools that would prompt its users to systematically map the institutional mechanisms and other concrete measures needed to help the process through. Power-leveling mechanisms need to be instituted at international, national, and subnational levels to ensure the accountability of the state and non-state actors, including international conservation groups such as the IUCN. This is especially relevant in the context of increasing militarization of nature protected areas, which has led to instances of gross violation of human rights, apparently under the watch and knowledge of some prominent global conservation NGOs.[15]

RPFF developed the HBF

The analyses presented here are based on the contents of RPFF, a memorandum document developed by the HBF based on extensive international consultations. While it is not a governance framework per se, the conceptual apparatus of 'resource politics' developed by the authors of RPFF sheds light on a number of questions at the center of the analyses presented in this chapter. The RPFF contents are focused on 'organizing use of natural resources' (HBF, 2014: 6) and on strategies 'to change who controls and uses nature in which way' (ibid.: 17). While a detailed analysis of the normative agenda of RPFF is beyond the scope of this chapter, an explicit attention to the patterns of resource control and use is an essential aspect of resource governance. Juxtaposing the RPFF's contents of resource politics to the key elements of the meta-framework developed here also points to the type of institutional interventions that would be needed for the realization of various goals of natural resources governance that the RPFF prioritizes.

Specification of goals

RPFF defines its goals as twofold: (a) to respect the ecological limits of our world and, at the same time, (b) strengthen human rights and democracy (HBF, 2014: 6). The contents of the memo are driven quite significantly by these specific goals. While many may see the second goal to be beyond the central focus of natural resource governance, an explicit statement of the RPFF's goals is noteworthy. This is crucial for a critical scrutiny of whether other elements of RPFF are designed to prioritize one or the other goal, although an actual assessment is beyond the scope of this chapter.

Representation of stakeholders

A democratic representation of key stakeholders is one of the central points of focus in RPFF. It advocates for 'more equity concerning who controls, who owns, who has access to and who decides over how humans and society interact with nature' (HBF, 2014: 37). As such, RPFF underlines the importance of equitable representation in all aspects of natural resource governance. Equally important, the writing of RPFF built on the following three tiers of deliberations over a period of two years (HBF, 2014: 9): (i) young people from 29 countries met in ten 'future workshops' held in all major world regions; (ii) delegates from each future workshop brought back insights and experiences into brainstorming sessions with the HBF team; and (iii) 11 internationally renowned environmentalists and human rights experts were part of an advisory board constituted for this purpose. In addition to espousing the importance of broad-based representation, the process of the development of RPFF demonstrates this in practice.

Monitoring and enforcement

RPFF advocates for 'direct regulatory measures like hard caps for CO_2 emissions and stricter environmental and social standards' (ibid.: 27), including proper regulations of financial markets (ibid.: 27) and monitoring of the environmental and social effects of global free trade (ibid.: 30). Equally important, despite a rather heavy reliance on international institutions and legal mechanisms, the memo argues for bringing 'popular pressure, mobilization and organization' to seek more effective enforcement of already existing international and national laws (ibid.: 42). This brings up important questions about the sources and the scope of governance arrangements, which are often dominated by elite policy makers.

Commitment of state power

RPFF authors hold the 'short-term interest' of governments responsible for enabling environmentally destructive practices at the hands of large private and state-owned corporations. Accordingly, the RPFF does not advocate for a significant role for the state in the pursuit of its agenda of resource politics. Instead, it pins its hope on global and international institutions (e.g., the UN bodies and conventions, multilateral development banks, or 'plurilateral government clubs' to support the goals of protecting the environment and the human rights of the marginalized) (ibid.: 40).

Powers and counterpowers

RPFF is focused on the goal of upsetting the concentration of power. It argues that private or state-owned enterprises are 'getting too big and too powerful to be governed…in most cases also too big to jail' (HBF, 2014: 28). They argue for more effective enforcement of global agreements on protection of human rights and ultimately relying on litigation to hold powerful actors to account (ibid.: 43). However, these well-articulated critiques do not translate into RPFF making concrete suggestions for institutional reforms or new institutional architecture meant to effectively distribute powers and counterpowers among key actors. The only exception to this is the suggestion about revisiting the global and international patent regimes to counterbalance the power that agroindustry corporations wield in controlling the production and propagation of seed varieties (ibid.: 44).

Analysis

The extent to which the three selected global frameworks for natural resource governance address the main elements of the meta-framework proposed here is summarized in Table 2.1.

The overview of the analysis of the three major frameworks for natural resource governance presented in Table 2.1 suggests that two of the three frameworks studied do not outline most of the five key dimensions of meta-governance identified above or do so only partially and ambiguously. The approach that comes closest to covering most of the dimensions of meta-governance proposed here is RPFF. It is noticeable that the HBF, which spearheaded the development of RPFF, is not invested directly in the business of resource governance. This reinforces the theoretical argument made in Section 1 to suggest that governance frameworks developed by actors and agencies invested in the governance of natural resources, including global think tanks, are likely to reflect their preferences, often in ways that are not made transparent.

TABLE 2.1 An assessment of natural resource governance frameworks based on the meta-framework

Key elements	*Resource Governance Index (RGI)*	*Natural Resource Governance Framework (NRGF)*	*Heinrich Böll Foundation (HBF)*
Specification of goals	Not explicitly	Not explicitly	Yes
Representation of stakeholders	Unclear	Ambiguous	Yes
Monitoring and enforcement	Selectively, in abstract	Ambiguous	Yes
Commitment of state power	No	No	No
Powers and counterpowers	Selectively, in abstract	Ambiguous	Partially

The NRGI caters to the needs of global and multinational corporations and national government actors interested in utilizing natural resource extraction as a source of state revenues. This is manifest in the RGI's focus on information-driven governance that serves the goals of private extraction of public resources. On the other hand, IUCN is invested in the protection of wildlife and biodiversity conservation, through the promotion of nature protected areas that are managed, by and large, along the lines of fortress conservation. Despite such a programmatic focus, the NRGF developed by the IUCN does not specifically ask how the goals of natural resource governance are to be prioritized and defined. Instead, NRGF focuses specifically on promoting a rights-based approach to conservation in which the rights of resource-dependent people are seen as an instrument for promoting nature conservation. Such an instrumentalist approach to conservation does not challenge the largely exclusionary nature of protected area model of conservation (Kashwan, 2013). The analysis presented above shows that the major governance blind spots were equally evident in the frameworks developed with two quite distinct mandates – the promotion of the corporate-led resource exploitation on the part of the NRGI and IUCN's promotion of the conservation of wildlife by excluding human intervention altogether. This is owed to the fact that in each of these cases, irrespective of the stated goals, the governance framework was motivated quite significantly by specific and narrowly defined normative goals. In each of these cases, the think tanks or the international agency involved should be seen as part of a specific interest group, not as neutral arbiters of natural resource governance.

The analysis of the IUCN-led NRGF is especially instructive. For an organization devoted to the agenda of environmental conservation, the NRGF does not explicitly recognize the importance of monitoring and enforcement in natural resource governance. This intriguing omission can be understood if NRGF is seen as a strategic intervention that is meant to quell the criticism that big international conservation NGOs have faced because of their apparent neglect of the rights of natural resource-dependent communities. In some cases, these NGOs have been accused of turning a blind eye to the environmental destruction caused by powerful militaries and national governments.[16] NRGF is clearly an attempt to address such criticism by coming up with documents that offer radical support for the rights without bringing about structural transformations in the dominant models of conservation. In this sense, NRGF mimics IUCN's advocacy of rights-based approaches to conservation and its joining of the Conservation Initiative on Human Rights (CIHR). Irrespective of the sincerity of intentions, each of these efforts has had the impact of creating a rather artificial separation between some of the core tasks of natural resource governance (e.g., monitoring, enforcement, and accountability) and their implications for questions of equity and justice (see Duffy, 2014; Kashwan, 2013).

The second important insight we get from a comparative reading of the RGI and NRGF is the disjuncture between the governance approaches rooted in a western liberal framework, in which the rule of law is taken for granted, and its applications to the problems of resource governance in the global South. The RGI's measurement of effective governance is informational or legalistic in its orientation; it does not present measurement of the effectiveness of the laws, rules, and regulations per se. As such, RGI fits in the currently popular transparency-based model of governance (e.g., regulation by revelation or regulation by disclosure mode) (see Gupta, 2010). Interestingly, NRGF also makes a reference to 'open and accessible information regarding the actions of relevant authorities' as an important source of accountability (Springer, 2016: 9). However, as discussed above, regulation and governance by revelation would be grossly inadequate in the context of the developing countries that often lack the institutional infrastructure required for making information measures count in meaningful ways.

The third important insight the analysis above reveals is the attention that each of these groups devotes to the issue of the persistent and enormous gaps between policy knowledge and implementation (HBF, 2014: 6). The authors of RGI assert that these gaps are most serious in programmatic areas that matter most: 'local impacts and subnational resource revenue sharing' (NRGI, 2017b: 15). Despite the recognition that many environmental harms are caused by 'poorly regulated companies extracting resources' (NRGI, 2017b: 16), such post-facto acknowledgments do not factor into the RGI scores, which inform decision-making around resource prospecting and resource exploitation. Departing from the RGI's focus on policy pronouncements, the NRGF accounts for the importance of recording the actual outcomes. However, as the discussion above suggests, NRGF does so without an attention to the ways in which powerful state and non-state actors, including international conservation NGOs, influence resource governance in the realm of wildlife conservation. This gap is a bit counterintuitive considering that IUCN efforts are meant to respond to criticisms that the policies and programs of nature conservation, promoted by IUCN and other big conservation NGOs, failed to protect the rights of indigenous people and other forest-dependent people (Colchester, 2004).

The memo developed by the HBF addresses this question more profoundly. While relying most heavily on international agreements and international human rights law, the memo also suggests the need for engineering popular mobilization to seek more effective enforcement of already existing laws (HBF, 2014: 42). Such an emphasis lends a greater coherence to the memo's focus on 'resource politics' and helps situate resource governance within the broader political and economic processes. Notwithstanding the importance of mass mobilization, expecting the resource-dependent groups to mobilize against powerful national governments and global corporations would impose potentially unrealistic burdens on the poorest people. This suggests the need for the institutionalization of basic governance framework within a broader framework of the rule of law, which takes the role of the state as the anchor of institutional integrity. This is an especially tricky dimension to understand for the experts and think tanks located in the global North, who can afford to take the fundamental institutional integrity of the state for granted. As the delegates from the Böll Foundation's

Future Workshop for East and Horn of Africa suggested, the state has the essential obligation of protecting 'all the citizens' rights to a fair share of their natural resources, rights to adequate compensation' for the losses incurred by resource exploitation activities (HBF, 2014: 42). Such a role of the state as the *institutional anchor* is, unfortunately, missed quite often in analyses of resource governance, including by the authors of RPFF.

The meta-framework developed here seeks to address an important paradox – of recommending a central institutional role for the state despite the currently powerful and often abusive behaviors of many states and state agencies – via a systematic analysis of the powers and counterpowers of the most powerful and the most vulnerable actors in natural resource governance. Such a systematic analysis of the balance of power among different actors and agencies invested in or affected by a resource governance regime is necessary for an assessment of a number of important features of natural resource governance, including accountability and the effectiveness of monitoring and enforcement of harms to the environmental and human rights of resource-dependent groups. A reasonably well-balanced distribution of powers is also crucial for fostering 'credible commitments' needed for underwriting any governance system, especially in the face of market failures and other externalities. As the authors of RPFF suggest, large private or state-owned enterprises have become 'too big to fail – and therefore in most cases also too big to jail' (HBF, 2014: 28). The recent work on the science of climate responsibility offers a good model for kick-starting a process of global accountability in the arena of natural resource governance. Researchers have shown that 50 investor-owned corporate emitters (e.g., BP, Chevron, ConocoPhillips, ExxonMobil, Peabody, Shell, and Total) contributed 'around 10% of the global average temperature increase and roughly 4% of global sea level rise' between the period 1980 and 2010 in which fossil fuel companies knew that their products would contribute to global warming.[17]

While various types of global and transnational efforts, including those recommended by the Böll Foundation, are useful, in the end, the rule of law cannot be strengthened without creating a fair and effective institutional mechanism backed by the authority of the state at the national and subnational levels. The authority of the state has to be part of a broader architecture of powers and counterpowers disbursed among a variety of state and non-state actors.

Conclusion

The popularity of networked governance or the governance beyond government has had the effect of throwing the baby with the bathwater. Despite the widely acknowledged shortcomings of these popular governance models, a wide-based rethinking of the concept and its institutional manifestation is not all too common. Some scholars attribute these omissions to the attraction of a 'technical approach' to deal with the messy topic of political institutions and interactions – like corruption, accountability, and leadership (Grindle, 2010: 6).

In this chapter, I have presented a parsimonious meta-framework that conceptualizes arguments focused specifically on governing natural resources in a context of entrenched political and economic inequalities. Instead of making a priori assumptions about the virtues of non-state/market actors and the pitfalls of state agencies or elected representatives, the PEI approach, which informed the analyses presented in these pages, emphasizes the importance of putting in place institutional arrangements to create strong incentives and minimizing perverse incentives. Moreover, instead of thinking the process of 'institutional design' as a technical task, it takes seriously the importance of political and economic contexts and adopts a broader understanding of the rule of law. The notion of 'steering' referred to at the top of this text is used to seek a balance of regulation and incentivizing of appropriate behavior. However, to avoid the pitfall of specific interests driving the governance narrative, the meta-framework developed here recognizes that governance is essentially a 'political' matter. Whether a society's natural resources are devoted to commercial harvesting, are put under the stewardship of resource-dependent groups for meeting local needs, or are set aside for the ostensible goal of wildlife and biodiversity conservation is determined through a political process anchored in political institutions specific to a society (Kashwan, 2017). Global institutions must work alongside domestic institutions, if they seek to influence the design and enforcement of governance mechanisms.

This is specifically true of the ongoing efforts to introduce market-based or fund-based payment for ecosystem services. The success of market-based incentives is highly context-dependent, which is why these programs cannot be assumed to have uniform effects across different contexts (Corbera et al., 2007; Kashwan, 2015). Without careful attention to theoretically relevant contextual factors, market-based programs could end up fueling further commercialization and result in grave exploitation of the resource base and resource-dependent groups (Dauvergne, 2016). The success of environmental conservation is also tied closely to the 'arrangements by which [the rules controlling resource use] are forged, enforced, and revised' (Brockington et al., 2018: 1257).

The configuration of five key dimensions of natural resource governance developed here are intended to be internally coherent and externally generalizable. Internal consistency refers to the extent to which each of the five dimensions is theoretically aligned to other dimensions in the meta-framework. For instance, the extent of substantive representation of key stakeholder (#2) is linked to the balance of power in a society (#5), which has significant influence over the effectiveness of monitoring and enforcement (#3). Each of these dimensions is also linked to the extent to which the state acts as an institutional anchor in the domain of natural resource governance (#4). The framework also asks for making explicit the goals of a particular governance mechanism (#1), which would allow a theoretically driven analysis of whether each of the other four governance elements is aligned to support the pursuit of specific governance goals. Because this meta-framework makes no assumptions about the organic development or de facto existence of any of the major institutional prerequisites of fair and effective governance, it can be applied to the analyses of governance in a wide variety of settings.

The interconnectedness of the five key dimensions of governance means that these elements are often contingent on the institutionalization of politics and economics in a society. Yet 'institutionalization' cannot be engineered in a short time horizon, which presents the knotty problem of how best to foster effective governance, especially when the goal is to facilitate the pursuits of multiple social and environmental goals. While there are no overarching theories that address this challenge specifically, comparative analyses show that the emergence of inclusive institutions is dependent greatly on the inclusiveness of political and policy processes related to the deliberations of the goals, design, and implementation of resource governance arrangements (Adger et al., 2005; Chhatre, 2008; Kashwan, 2017). The deeply political nature of institutions presents a rather tricky challenge that cannot be addressed with one-shot institutional reforms, even if such reforms are multifaceted and intended to be transformational. Yet the solution to the paradox of the politics of governance is not to reduce this to a technical policy advice of the type rendered by NRGI or a sympathetic discourse as presented by the IUCN but instead to recognize the inevitability of politics. The meta-framework presented here offers a way of accounting for the politics and practices of resource governance arrangements.

Notes

1 A Google Scholar search with the phrase 'who is accountable to whom for what' brings up a number of articles, most of which are focused on specific sectors.
2 The framework has been widely used for institutional analysis related to common pool resources, with a variety of property regimes, but also public policies in arenas as diverse as policing, health services, and other public services of various types in both global North and South (for an overview of the application of IAD in varied fields, see Ostrom, 2007).
3 https://stats.oecd.org/glossary/detail.asp?ID=3215.
4 www.bbc.com/future/story/20150402-the-worst-place-on-earth
5 www.vox.com/2018/6/22/17493414/trump-executive-order-deepwater-horizon-drilling-oceans.
6 Similar omission of questions of power and inequality have also been identified in the related field of institutional analysis (e.g., in Elinor Ostrom's design principles for governing the commons and other works inspired by them) (Ostrom, 1990; Stern, 2011). However, a number of recent contributions have addressed these gaps in the theories of institutional analysis (Clement, 2010; Kashwan, 2016; Kashwan et al., 2018). We apply those insights for the development of the meta-framework proposed here.
7 For two prominent commentaries on how the questions of power and power asymmetries have also been neglected in the related field of institutional analysis, see Knight (1992) and Moe (2005). For a recent overview, see Kashwan et al. (2018).
8 One must note that such uneven measurement of subcomponents makes RGI unsuitable for any kind of cross-statistical analysis.
9 https://info.worldbank.org/governance/wgi/pdf/rq.pdf
10 'The average score for laws governing the local impacts of extraction is 64 points – but countries score only 23 points for the application of these laws. This usually results from countries' failure to implement environmental regulations.' (NRGI, 2017b: 15).

11 The full set of questionnaires can be accessed here: https://resourcegovernanceindex.org/system/documents/documents/000/000/075/original/2017_RGI_Questionnaire.xlsx?1498601420
12 For a critique of such instrumentalist approaches, see Kashwan (2013).
13 www.iucn.org/commissions/commission-environmental-economic-and-social-policy/our-work/knowledge-baskets/natural-resource-governance
14 See, for instance, Statement of Ms. Victoria Tauli-Corpuz, Special Rapporteur on the Rights of Indigenous Peoples, at the 17th Session of the United Nations Permanent Forum on Indigenous Issues. https://ohchr.org/EN/NewsEvents/Pages/DisplayNews.aspx?NewsID=22961&LangID=E
15 'A Leaked Report Shows WWF Was Warned Years Ago Of "Frightening" Abuses' www.buzzfeednews.com/article/katiejmbaker/wwf-report-human-rights-abuses-rangers
16 www.thenation.com/article/front-lines-new-pacific-war/; https://www.mondialisation.ca/devastating-environmental-impacts-greenwashing-wars-and-the-us-military/5547393
17 www.ucsusa.org/fa17-ideas-in-action#.XGmKVqBOl6o

References

Adger, W. N., Brown, K. and Tompkins, E. L. (2005) 'The political economy of cross-scale networks in resource co-management', *Ecology and Society*, vol. 10, no. 2: 9. http://www.ecologyandsociety.org/vol10/iss2/art9/.

Agrawal, A. and Ribot, J. C. (1999) 'Accountability in decentralization: A framework with South Asian and West African Cases', *The Journal of Developing Areas*, vol. 33, no. 4, pp. 73–502.

Andrews, M. (2008) 'The good governance agenda: Beyond indicators without theory', *Oxford Development Studies*, vol. 36, no. 4, pp. 379–407.

Bank, W. (1994) *Governance: The World Bank's Experience*. World Bank, Washington, DC.

Bardhan, P. (2002) 'Decentralization of governance and development', *Journal of Economic Perspectives*, vol. 16, no. 4, pp. 185–205.

Boltanski, L. and Chiapello, E. (2005) *The New Spirit of Capitalism*. Verso Books, London and New York.

Brockington, D. (2002) *Fortress Conservation: The Preservation of the Mkomazi Game Reserve, Tanzania*. International African Institute in Association with Indiana University Press, Bloomington.

Brockington, D., Adams, W. M., Agarwal, B., Agrawal, A., Büscher, B., Chhatre, A., Duffy, R., Fletcher, R. and Oldekop, J. A. (2018) 'Working governance for working land', *Science*, vol. 362, no. 6420, pp. 1257–1257.

Campese, J., Nakangu, B., Silverman, A. and Springer, J. (2016) *The NRGF Assessment Guide: Learning for Improved Natural Resource Governance*. NRGF Paper. IUCN and CEESP, Gland.

Chhatre, A. (2008) 'Political articulation and accountability in decentralisation: Theory and evidence from India', *Conservation and Society*, vol. 6, no. 1, pp. 12–23.

Clement, F. (2010) 'Analysing decentralised natural resource governance: Proposition for a "politicised" institutional analysis and development framework', *Policy Sciences*, vol. 43, no. 2, pp. 129–156.

Colchester, M. (2004) 'Conservation policy and indigenous peoples', *Environmental Science and Policy*, vol. 7, no. 3, pp. 145–153.

Corbera, E., Brown, K. and Adger, W. N. (2007) 'The equity and legitimacy of markets for ecosystem services', *Development and Change*, vol. 38, no. 4, pp. 587–613.

Crawford, A. (2006) 'Networked governance and the post-regulatory state? Steering, rowing and anchoring the provision of policing and security', *Theoretical Criminology*, vol. 10, no. 4, pp. 449–479.

Dauvergne, P. (2016) *Environmentalism of the Rich*. MIT Press, Cambridge, MA.
Davies, J. S. (2011) *Challenging Governance Theory: From Networks to Hegemony*, 1st edition. Bristol University Press, Bristol.
Dressler, W. and Roth, R. (2011) 'The good, the bad, and the contradictory: Neoliberal conservation governance in rural Southeast Asia', *World Development*, vol. 39, no. 5, pp. 851–862.
Duffy, R. (2014) 'Waging a war to save biodiversity: The rise of militarized conservation', *International Affairs*, vol. 90, no. 4, pp. 819–834.
Duffy, R. (2016) 'War, by conservation', *Geoforum*, vol. 69, pp. 238–248.
Gisselquist, R. (2012) *Good Governance as a Concept, and Why this Matters for Development Policy*. World Institute for Development Economics Research, United Nations University, Helsinki.
Grindle, M. (2010) 'Good governance: The inflation of an idea', *HKS Faculty Research Working Paper Series, RWP10-023*.
Gupta, A. (2010) 'Transparency in global environmental governance: A coming of age?', *Global Environmental Politics*, vol. 10, no. 3, pp. 1–9.
Heinrich Böll Foundation (HBF) (ed). (2014) *Resource Politics for a Fair Future: A Memorandum of the Heinrich Böll Foundation*. Heinrich Böll Foundation, Berlin.
Hurwicz, L. (2008) 'But who will guard the guardians?', *American Economic Review*, vol. 98, no. 3, pp. 577–585.
Kashwan, P. (2013) 'The politics of rights-based approaches in conservation', *Land Use Policy*, vol. 31, pp. 613–626.
Kashwan, P. (2015) 'Forest policy, institutions, and REDD+ in India, Tanzania, and Mexico', *Global Environmental Politics*, vol. 15, no. 3, pp. 95–117.
Kashwan, P. (2016) 'Integrating power in institutional analysis: A micro-foundation perspective', *Journal of Theoretical Politics*, vol. 28, no. 1, pp. 5–26.
Kashwan, P. (2017) *Democracy in the Woods: Environmental Conservation and Social Justice in India, Tanzania, and Mexico*. Oxford University Press, New York.
Kashwan, P., MacLean, L. M. and García-López, G. A. (2018) 'Rethinking power and institutions in the shadows of neoliberalism: (An introduction to a special issue of World Development)' *World Development*, vol. 120, pp. 133–146.
Knight, J. (1992) *Institutions and Social Conflict*. Cambridge University Press, New York.
Kolstad, I. and Søreide, T. (2009) 'Corruption in natural resource management: Implications for policy makers', *Resources Policy*, vol. 34, no. 4, pp. 214–226.
Lockwood, M., Davidson, J., Curtis, A., Stratford, E. and Griffith, R. (2010) 'Governance principles for natural resource management', *Society and Natural Resources*, vol. 23, no. 10, pp. 986–1001.
Michener, G. (2015) 'Policy evaluation via composite indexes: Qualitative lessons from international transparency policy indexes', *World Development*, vol. 74, pp. 184–196.
Moe, T. M. (2005) 'Power and political institutions', *Perspectives on Politics*, vol. 3, no. 2, 215.
NRGI (2014) *Natural Resource Charter*, 2nd edition. Natural Resource Governance Institute, Oxford.
NRGI (2017a) *2017 Resource Governance Index: Method Paper*. Natural Resource Governance Institute, Oxford.
NRGI (2017b) *2017 Resource Governance Index*. Natural Resource Governance Institute, Oxford.
Nunan, F. (2006) 'Empowerment and institutions: Managing fisheries in Uganda', *World Development*, vol. 34, no. 7, pp. 1316–1332.

Ostrom, E. (1990) *Governing the Commons: The Evolution of Institutions for Collective Action*. Cambridge University Press, New York.

Ostrom, E. (2007) 'Institutional rational choice: An assessment of the institutional analysis and development framework', in P. A. Sabatier (ed.), *Theories of the Policy Process*. Westview Press, Boulder, CO, pp. 35–71.

Rhodes, R. A. W. (1996) 'The new governance: Governing without government', *Political Studies*, vol. 44, no. 4, pp. 652–667.

Ribot, J. C., Agrawal, A. and Larson, A. M. (2006) 'Recentralizing while decentralizing: How national governments reappropriate forest resources', *World Development*, vol. 34, no. 11, pp. 1864–1886.

Schlager, E. (2007) 'A comparison of frameworks, theories, and models of policy processes', in P. Sabatier (ed.), *Theories of the Policy Process*. Westview Press, Boulder, CO.

Springer, J. (2016) *Initial Design Document for a Natural Resource Governance Framework*. NRGF Working Paper No. 1. IUCN and CEESP, Gland.

Stern, P. (2011) 'Design principles for global commons: Natural resources and emerging technologies', *International Journal of the Commons*, vol. 5, no. 2, pp. 213–232.

Stoker, G. (2011) 'Was local governance such a good idea? A global comparative perspective', *Public Administration*, vol. 89, no. 1, pp. 15–31.

Vatn, A. and Bromley, D. W. (1997) 'Externalities—A market model failure', *Environmental and Resource Economics*, vol. 9, no. 2, pp. 135–151.

World Bank (1989) *From Crisis to Sustainable Growth-Sub-Saharan Africa: A Long-Term Perspective Study*. The International Bank for Reconstruction and Development/The World Bank, Washington, DC.

World Bank (2019) Governance Global Practice. www.worldbank.org/en/topic/governance/overview#2 Accessed 18/02/2019.

3

NETWORKED PARTICIPATION

How social network analysis can inform participatory processes in environmental governance

María Mancilla García and Örjan Bodin

Introduction

The governance of natural resources is characterized by the presence of actors with different interests over the use of resources. Simply speaking, actors either compete for the access to the resource, or they cooperate to devise rules that facilitate resource sharing without detrimental overuse (Ostrom, 1990). Often, however, there is a mixture of cooperative and competitive processes acting in parallel, such as when actors are seeking alliances to defend their interests towards competitors (Sabatier, 1988; Sabatier and Jenkins-Smith, 1999; Weible and Sabatier, 2009), or when cooperative and competitive social ties bind together the very same set of actors (Basurto et al., 2016; Robins et al., 2011). Social network analysis (SNA) is a perspective that focuses on actors and their social relationships, and as such has been found very useful to better understand how different social and governing structures emerge and evolve (Ingold and Fischer, 2014) or how certain social network structures seemingly give rise to different social and ecological outcomes (Barnes et al., 2016). SNA has been extensively used to study governance, and numerous scholars have synthesized the state of the field (see for example Bodin, 2017; Henry et al., 2012; Lubell et al., 2012; Nowell et al., 2018; Robins et al., 2012; Scott and Ulibarri, 2019). In the natural resource governance literature, network approaches are increasingly used to study different forms of cooperation, collaboration, and collective action (Alcañiz and Berardo, 2016; Angst et al., 2018; McAllister et al., 2017). This chapter focuses, in particular, on the study of participatory forums for natural resource governance.

Participatory models have been introduced all over the world as a way to initiate debate between actors holding different perspectives and interests. These vary from one-time consultation procedures to permanent participatory forums in which a broad set of stakeholders are brought together on a regular basis. The empirical literature on participation has raised concerns about the ability of these diverse forms of participatory *governance* to effectively include previously excluded voices. Among the main problems identified are (i) whether participants actually represent collective interests instead of their own personal interests, (ii) power asymmetries among participants leading to varying levels of influence and impact, and (iii) the existence of coalitions defined by their members sharing common interests, which can prevent open and deliberate dialogues involving all participants.

In order to ensure that participatory forums effectively fulfill their mission of including different voices in governance and allowing for open and deliberative dialogues, it is crucial to understand how actors relate to each other when joining and engaging in such forums. We argue that many of the issues that the literature identifies as key to evaluating participatory forums can be investigated and analyzed from a social network perspective. SNA allows the analyst to consider the specific context of each participatory forum (e.g., identifying the most powerful actors in a specific context) while making insights comparable across contexts (e.g., what specific attributes of participating actors tend to coincide with the establishment of mutual social ties). Network analysis focuses on patterns of social relations and is therefore particularly suitable for the investigation of reoccurring social dynamics in participatory governance models (such as participatory forums), where relations are maintained and developed through time.

In this chapter, we briefly discuss the key concepts and questions in the literature on participatory governance and show how network analysis can help us investigate them. We then briefly present the methodological and conceptual first steps necessary to engage in network analysis and introduce exponential random graph models (ERGMs), a modeling technique that allows the analyst to investigate the empirical network by discerning the prevalence (or lack of) specific network configurations ('building blocks'). We take a case of a participatory forum for water governance in Brazil as an example to illustrate how ERGMs can assist the investigation of collaboration networks. We then interpret the results to identify implications for the ability of the participatory forum studied to achieve its objectives.

Participation in environmental governance: theoretical perspectives

Participation is a broad term that refers to governance modes that range from isolated public consultations to permanent deliberative collegiate forums of governance (Newig et al., 2018). We will focus here on the latter, since it is in such forums that participating actors are required to collaborate. The difference between the concepts of participation and collaboration is sometimes blurry. We follow Newig's distinction, according to which participation is a term that insists on the inclusion of actors who did not previously take part in management, while the term collaboration focuses on the process of obtaining joint decisions (Newig et al., 2018). Yet we recognize that the concepts are linked and that collaboration is one of the activities that occurs in participatory forums – others can be, for example, conflict, learning, or information exchange.

Participatory modes of governance have received sustained attention in the scholarly literature as they have progressively been implemented across an increasingly large number of environmental governance areas and countries. Besides the traditional justification of participation as a means to increase legitimacy in public management, in recent years the 'instrumental rationale' (Newig and Kvarda, 2012) has also been appealed to, arguing that participatory governance allows for integration of numerous perspectives (Edelenbos et al., 2011) and facilitates the implementation of the decisions made (Bulkeley and Mol, 2014). Yet the capacity of participatory governance to actually produce better environmental outcomes remains uncertain (Beierle and Cayford, 2002; Newig et al., 2018; Young et al., 2013). Most of the literature acknowledges that the benefits of participatory environmental governance, both in terms of democratic purposes and effective environmental outcomes, are dependent on contextual factors (Mancilla García and Bodin, 2019; Irvin and Stansbury, 2004). We demonstrate that network analysis provides a good entry point to investigate these issues across contexts.

Like 'participation,' 'network' is also a term that has been defined in different ways in the environmental governance literature. Networks can be used as a method to analyze a particular governance setting (Leifeld, 2013), as a metaphor to describe a particular reality such as a group of connected actors (also called implicit network studies (Bodin and Prell, 2011; Scott and Ulibarri, 2019)), or as a theoretical framework (such as policy networks (Ingold and Leifeld, 2016)). We focus here on SNA. It has been extensively discussed whether SNA constitutes simply a method or more than that, that is, a framework for social analysis (Freeman, 2004). Since this debate is beyond the focus of this chapter, we set it aside and adopt a position acknowledging both sides. Indeed, we argue that network analysis is a method that comes with an explicit perspective: investigating a set of actors and the relations among them. This perspective can then be used to theorize if and why relationships are formed, and what effect and outcomes this can lead to. But it can also be used just to map out relationships thereby simply describing the social relational structure of a group of actors at a given point in time.

When studying the relations among actors taking part in any given governance setting, the activities typically considered are resource exchange, learning and collaboration (Lubell et al., 2017). Network analysis allows for the study of the flows of a given resource, such as information. By mapping the network of a set of connected actors – such as participants in a water basin forum – one can investigate how information flows among those actors, or beyond them, investigating each actor's own network within and beyond the given forum. We can also study whom is the most sought-after actor to obtain information from. Additionally, the identification of clusters can signify that groups that exist outside of the participatory forums are reproduced in the participatory setting. Identifying actors that connect clusters can help us understand which actors are key in ensuring cooperation between groups.

Finally, the literature on participatory governance has long investigated the attributes that explain why certain actors are more important than others. Three attributes have been identified as being particularly important: preference similarity, reputational power and formal authority (Fischer and Sciarini, 2016). Some studies have found that actors tend to collaborate with those who are similar to them (Henry, 2011). Other studies have shown that actors link up with those perceived as most powerful (Weible, 2005), or with those having formal authority (König and Bräuninger, 1998). This perspective holds that actors link with others for strategic reasons rather than as a result of belief similarity. In the next section, we further investigate these issues with a special focus on the study of participatory dynamics.

Network analysis for participatory forums: first steps

The first task of researchers willing to engage in SNA is typically to determine the population of interest. In terms of environmental resource governance, this could be the population of stakeholders with interest in a resource, and/or those actively participating in its governance. If we take the case of water governance, basin-based participatory forums for water governance are a paradigmatic example of a group of actors with different interests, actively participating in the governance of the resource. In network analysis each of those actors would be called a 'node.' Although we speak here of 'actors' for the sake of simplicity, nodes can also be organizations, or forums, or more abstract elements such as tasks or problems. In other words, nodes are the units of analysis relevant for a given governance setting.

The population of interest for a given study can be identified in different ways. In some cases, the analyst can use lists of membership of a particular forum, creating a socio-centric network, that is, a network of a given system. A network that exists as a result of a shared working space is called a serendipitous network (Scott and Ulibarri, 2019). In other cases, the population of relevance might be largely unknown, and thus defining it becomes a research endeavor in its own right. In such cases, one can engage in snowball sampling, selecting a first small sample of actors who are certainly involved in the problem at hand, and expanding the sample based on the responses that such actors provide on who else is involved. In that way, the analyst can construct many egocentric networks, in which what is investigated are the relationships a particular actor – ego – has with others – alters. Furthermore, the snowball sampling can result in the identification of a bounded set of actors, that is, at some point further sampling does not reveal any new actors. This group of actors can then be defined to constitute a 'goal-oriented' network in which participants come together for a purpose, and it is the fulfillment of that purpose that maintains the network.

Once the population is determined, the subsequent question to ask is what are the relations at play between the nodes, that is, how the nodes interact (Scott and Ulibarri, 2019). The network is therefore made of nodes, their relations (also called 'edges,' 'links,' or 'ties') and the activity they perform – such as information exchange, collaboration, and learning. Different network configurations provide insights into actors' behavior and the type of network they produce. For example, if we observe a very dense network, this suggests numerous reciprocated links, which makes monitoring of defection easy and likely, and increases trust among network members (Berardo and Scholz, 2010). Additionally, the links that exist in a network can be more or less strong, depending on the frequency and/or intensity of the activity undertaken. These measures help to determine which actors are better connected and whether we can identify clusters of actors that might keep certain others apart.

These phenomena can also be studied as they evolve over time, for which longitudinal network data is necessary. The literature has argued that jointly participating in the same governance forums provides an opportunity for participants to see each other, recurrently facilitating communication, which might in turn enhance trust and joint learning (Lubell, 2007). Through SNA we can investigate whether, with time, participants in these forums increase the number of partners with whom they exchange information, for example as a result of building trust.

To sum up, through SNA one can investigate (i) what a particular network looks like, that is, what type of connections there are – which is called a structurally descriptive network study – and (ii) the reasons why a network has a particular structure, which can be called a structurally explicit network study (Alexander and Armitage, 2015; Scott and Ulibarri, 2019).

Stochastic modeling

While descriptive network statistics helps to reveal what a network looks like, they do not provide explanation for why the network looks the way it does nor for the behaviors that explain how people link to each other. ERGM are commonly used in the analysis of networks to investigate what determines particular structural characteristics, such as, for example, what factors explain the tendency actors have to link to others. ERGMs are built on likelihood statistics and allow for the investigation of the prevalence of certain building blocks (configurations) to be present in the network (Box 3.1). Such configurations (also called motifs, see Milo et al., 2002) can be linked to diverse theories and processes, such as dynamics of trust, risk aversion, strategic alliances, or reciprocity (Robins, 2015). ERGMs allow for the testing of several of those configurations simultaneously as well as the effect of a diverse set of attributes and thus are particularly useful for theory development (Lusher et al., 2013). However, without longitudinal data, the link from structure to process is inferred, not explicitly measured.

ERGMs are a type of model specifically designed for network analysis, that is, considering the various interdependencies of network data. The main difference between statistical methods such as regression and ERGM is that statistical methods assume independence of data observations while network analysis assumes data points are interdependent. A network is indeed created through patterns of connections and thus implies interdependence of data points.

BOX 3.1 TECHNICAL SPECIFICITIES OF EXPONENTIAL RANDOM GRAPH MODELS

ERGM uses maximum likelihood simulation techniques to fit a parameter vector Θ to a stochastic network model (Lusher et al., 2013):

$$P^{\theta}(X = x) \propto \exp\left\{\theta s(x)\right\} \qquad (1)$$

where X is a random network (x is the empirical network) and s(x) is a known vector of graph statistics (configurations) on x.

A well-fitting ERGM would then be able to adequately represent the network as being the result of a set of configurations and their associated coefficients. The coefficients capture if a certain configuration is suppressed or enhanced. Yet fitting an ERGM can suffer from convergence problems. Goodness-of-fit tests investigate whether the model fits the data, that is, these tests give the analyst the means to confront the selected configurations and their estimated parameters with many other possible configurations. The results of such tests indicate if the model is suffering from convergence problems, and if so what tendencies in the data are being missed (i.e., which other configurations, besides the explicitly modeled ones, are not well captured by the ERGM).

To translate an ERGM into regression terms, we could say that ERGMs conceive the network as the dependent variable, to be explained through different micro-level configurations or attributes, which would be the independent variables. These micro-configurations or motifs arise as a result of actors' behavior. ERGMs allow for the testing of whether certain ways of interaction better explain the observed network than others. ERGMs indicate the probability of a given micro-structure or attribute to be of relevance when explaining the overall network structure, in comparison to all possible structures. ERGMs allow for the simultaneous testing of several configurations at the same time and comparison enables the identification of which are more significant in explaining the structure of a given network.

Formulation of hypotheses related to participation

Based on our theoretical interest in participatory dynamics, we determined a set of hypotheses on which elements could help us explain why certain actors are more central than others in the collaboration network in a participatory water basin council:

Hypothesis 1: The more involved actors are, the more central they are

The literature on participatory governance has shown through diverse empirical studies that time availability and willingness to engage are fundamental for the functioning of participatory forums. Conversely, lack of time and logistical difficulties in participating (such as long travel distances) are identified parameters preventing involvement and perpetuating exclusion (Altschuler and Corrales, 2012; Davies and White, 2012; Diduck and Sinclair, 2002; Lund and Saito-Jensen, 2013).

Hypothesis 2: The longer actors have participated in the council, the more central they are

Participating over long time frames has been associated with the acquisition of the necessary knowledge to meaningfully participate (Sandström et al., 2014) and communicate (Jacobs et al., 2010). Some of the participants are active in participatory forums for longer than others, and inspired by the literature, we hypothesized that the participants that had spent the longest time participating in the forum had acquired institutional knowledge – provided by long-term experience – that would make them appear as interesting to link up to in the eyes of others.

Hypothesis 3: The more forums actors participate in, the more central they are

Recent literature on the Ecology of Games has argued that in order to understand the governance network in a particular system it is important to consider the effect of different forums in the system (Lubell et al., 2010) and on each other (Mancilla García and Bodin, 2018). We hypothesize that those actors who participate in several other forums would be perceived as having more influence on the system – by having access to more actors, gaining new knowledge or developing a global view on the system – than those who participate in fewer, and therefore would be more sought-after for collaboration. This is a debated matter since participation across forums takes time, and therefore those who participate in many forums might be perceived as dispersed.

Hypothesis 4: The more influential actors are perceived to be, the more central they are

Influential power, that is, 'the ability of individuals or groups to influence the agenda and final policy choice' (Akbulut and Soylu, 2012: 1146) is unequally distributed among participants. Some actors are better placed to determine the decisions made in participatory processes, where power dynamics are at play (Barnaud and Paassen, 2013; Cornwall and Coelho, 2006; Singleton, 2000). Hypothesis 4 considers that the level of perceived influence of an actor is key in determining others' decision to link up with them. This hypothesis assumes that actors align their position with others for strategic reasons (those more influential are more likely to get things done) rather than as a result of other variables, such as belief similarity.

Hypothesis 5: The more popular actors are, the more central they are

The tendency of those who are popular to become ever more popular is a basic social structure (Robins, 2015). Including this hypothesis to investigate collaboration dynamics could signal that the network is structured around certain actors although those don't necessarily hold the particular attributes selected, such actors might happen to be popular in the region studied or might each present unique characteristics that cannot be captured through attributes.

Hypothesis 6: Sector-based coalitions are reproduced in participatory forums

Coalitions can be defined in different ways; we adopt a definition inspired by the Advocacy Coalition Framework (ACF) (Sabatier, 1988; Sabatier and Jenkins Smith, 1999). We consider coalitions as groups of actors who share common beliefs and interests and act based on those interests. We hypothesized that coordination patterns corresponded to deeply rooted differences and/or conflicts of interests between sectors, inspired by the literature that discusses government entities or large resourceful private actors as holders of power (Dahl, 1994; Furlong, 1997; Yackee and Yackee, 2006).

Hypothesis 7: Actors tend to reciprocate links

Reciprocation is a common social tendency in networks. It is associated with bonding structures that increase trust and commitment (Berardo and Scholz, 2010; Lubell et al., 2012). We included this hypothesis to assess whether actors tend to work together, suggesting horizontality in relations, rather than follow a leader.

Table 3.1 provides a summary of the questions that we were interested in testing as they emerged from the literature on participatory governance, and the network approach followed to test them.

The case: basin council of 'Médio Paraíba do Sul' in Brazil

To evaluate the hypotheses discussed above, we focus on a water basin council with jurisdiction over part of the Paraíba do Sul river basin in Brazil, the so-called 'Médio Paraíba do Sul,' an area within the state of Rio de Janeiro (see Figure 3.1). The Brazilian legislation on water resources was approved in 1997, through law 9433. This law established the National Water Resource Management System which included the creation of water basin councils and of water agencies, taking inspiration from the French model of management and building on the basis of the 1988 Brazilian Constitution, which established two fundamental principles: water is a public asset, and the jurisdiction over it is shared

TABLE 3.1 Research questions and network approaches

Research question	*Theoretical insight*	*Network analysis approach*	*Configuration*	*Configuration name*
Who is central?	Determining who is a central actor provides information on which actors receive most links from others, that is, who are the actors that others want to link to. The information that centrality provides is dependent on what activity constitutes the network. For example, people who are central in terms of information exchange, signal people who are perceived as knowledgeable by others. Considering that one of the main research questions in the study of participatory research forums is whether or not participatory forums effectively manage to make previously excluded actors heard, an important question is whether actors who were previously excluded from management – such as civil society – appear among the most central actors in a network of collaboration, that is, whether all actors – irrespective of their background – seek to collaborate with actors from previously excluded backgrounds, such as civil society.	Centrality is the most common descriptive statistic in network analysis and there are many different types of centrality (see Robins, 2015). We focus here on 'degree centrality,' that is, a measure that indicates centrality in terms of the number of ties an actor has. There are two possible types of degree centrality – in-degree and out-degree centrality. In-degree centrality is calculated by counting the number of ties that a node receives, and out-degree by counting the number of ties that a node sends to others. Network analysis software allows for the visualization of centrality measures by, for example, defining the most central nodes bigger. To determine whether particular actors appear as the most central actors, nodes can be given different shapes depending on different attributes (see Figure 3.2).	Network level	Centrality

Why are people central?	A number of reasons can explain why certain people are more central than others, as we have seen with the hypotheses above.	We determined four attributes: Degree of involvement, Time in the council, Number of other forums attended and Average perceived influence. We tested through ERGM whether actors with those attributes were more sought-after than those without.		X_Receiver, alter effect
		We additionally studied through ERGMs whether actors tended to collaborate with those who are already central, that is, whether popularity drives centrality.	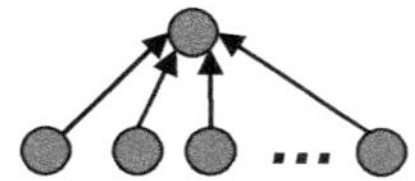	AinS
Do people link to those similar to them?	The literature on participation has investigated whether coalitions of interests that exist outside of participatory forums are simply reproduced within them or whether the forums encourage actors to exchange beyond their immediate friends. Since the objective of participatory forums is to bring actors together so that they can link beyond the group they belong to, too much homogeneity could be seen as failing that objective.	In network analysis this is generally referred to as 'homophily' – that is, the concept that actors like to link to those that are like them, those who share an attribute with them. In this case, we focused on the 'sector' to which participants belonged following the hypothesis that actors would constitute sector-based coalitions.		X_Match
Do people tend to reciprocate links?	When it comes to using this configuration to understand participatory dynamics, if reciprocity happens across the network, it would suggest that actors trust others regardless of their attributes.	This is the most basic ERGM configuration, as it is considered a normal social tendency, and is typically included in all studies. An incoming tie from an actor is reciprocated with an outgoing tie to the same actor. This for example means that if actor A reports aligning positions with actor B, actor B confirms it.		Reciprocity

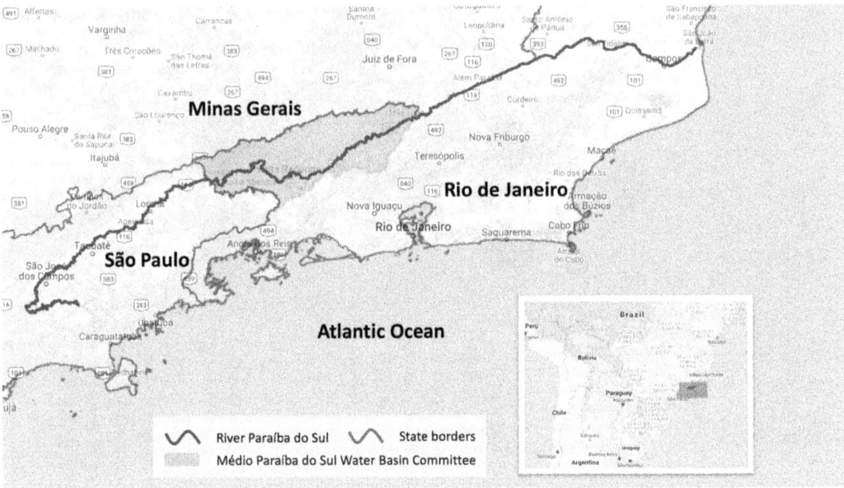

FIGURE 3.1 Case study area (Basemap Google Maps).

between the states and the federal level. Additionally, the law created Water Resource Committees at the state and national levels and their executive branch, State Water Resource Management Entities. It was only in the year 2000 that the National Water Agency was created, with the responsibility of implementing water policy instruments at the national level. The Brazilian agency is part of the Ministry of the Environment, and the minister of the environment is the president of the National Water Resource Committee, the highest organization in terms of water management at the federal level.

Water Basin Councils in Brazil are collegiate organisms that follow a tripartite design, in which representatives from the public sector – which includes the state and municipal governments for those basins with limits within a single state, and additional representatives from the federal government for those basins that cross several states – the private sector and civil society constitute the plenary. The exact composition of the council plenary is determined by each of the councils, the only condition the law establishes is that government entities representatives do not surpass half of the total participants. Councils also have one or several technical chambers – composed of two representatives of each sector – with several working groups and a directorate. In the case of the Médio Paraíba, the plenary elects a six-member directorate for two years with two representatives from each sector.

Brazilian councils are inserted within a web of participatory organizations for water management. The water basin council of the Médio Paraíba do Sul is subject to the legislation in force within the state of Rio de Janeiro, and it is part of a larger basin that includes the states of São Paulo and Minas Gerais. A federal council has jurisdiction over the entire River Paraíba do Sul basin. The Comitê do Médio Paraíba do Sul, of interest here, was set up in 2008 and brings together eight representatives of government entities, eight representatives of civil society and eight representatives of water users. At the time of fieldwork, seven of the government entities seats were assigned to different municipalities whereas one actor represented the state government of Rio de Janeiro. Representatives from civil society included different associations active in the region as well as the technical university in one of the main cities of the region, Volta Redonda. Users included sewerage treatment companies, industrial companies, sand extraction companies and representatives of the industrial federation of the state.

Data collection methods

The sample included in our study consists of the participants in the water basin council Médio Paraíba plenary, provided that they fulfilled one condition: having attended at least two of the last six plenary meetings the council had held. At the time of fieldwork, one of the civil society seats was vacant, and two participants did not fulfill the sampling condition. This resulted in a sample of 21 people of which one refused to participate in the study.

We presented respondents with the list of participants in the plenary of the council and asked them to evaluate each participant in terms of whether they coordinated positions with them – in terms of voting behavior and support in agenda setting and discussions – and whether they saw them as influential. The first question was rated on a four-point scale where 1 means 'rarely or never,' 2 means 'sometimes,' 3 means 'often,' and 4 means 'constantly.' For example, if participants coordinated on how to vote on every decision with another participant, they would give them a 4. With this data we constructed the collaboration network.

The question on influence used a five-point scale where 1 was 'not at all' and 5 'completely.' For both scales, 0 indicated an unknown participant. We use influence perception as a measure of reputational power; perceived influence was assessed based on the average rating from all other actors that indicated that they knew about an actor (attribute *influence*). As Fischer and Sciarini pinpoint,

> reputational power is admittedly subjective and may thus lead to inaccurate evaluations. But even if an evaluation is inaccurate, the simple fact that ego believes that alter is powerful will lead ego to behave as if alter were powerful, which will indeed empower alter ('self-fulfilling prophecy').
>
> *(2016: 181)*

Additionally, we asked actors a series of descriptive questions about their participation in the water management system that allowed us to develop a set of attributes and variables to further investigate actors' behavior and strategies. We asked participants to specify the number of years they had spent participating in the council (attribute *time*). We also asked them to list other forums in which they participated to discuss water management issues (attribute *forums*). The most common among these included other councils at the basin level and the state-based committee on water resource management. In less common cases, other forums were mentioned such as the national council on water resource management, a forum set up by several NGOs to discuss issues related to water quality and so on. Finally, we asked participants the degree to which they were involved in the council, which they self-assessed on a 1 to 4 scale where 1 is 'not at all' and 4 is 'a fundamental aspect of my work' (attribute *involvement*).

Collaboration network in the studied basin

As explained above, we focused for this study on the collaboration network among participants in the Comitê Médio Paraíba water council. We used a non-weighted network of collaboration constructed by focusing only on strong ties, that is, ties of strength 3 and 4. In order to visualize the main characteristics of this network, we used Net-Draw, an add-on to the Uci-Net software (Figure 3.2). We chose to assign a particular shape to nodes based on their sector, that is, whether they were representatives from government entities (circles), civil society (square), or the user sector (triangle). For reasons of confidentiality we cannot go into further details.

We chose to determine the size of each node based on its in-degree centrality, that is, on the number of ties a given node receives. These are calculated as the result of people's responses to the question on with whom they aligned positions. We chose this measurement instead of self-assessment, that is, the number of out-going ties that actors report since it is not uncommon that certain individuals exaggerate – intentionally or unintentionally – their number of social ties. Therefore, the aggregate assessment by others is a less-biased measure of centrality. As a result, our network is a directed network, that is, a network where the directionality of the links matters.

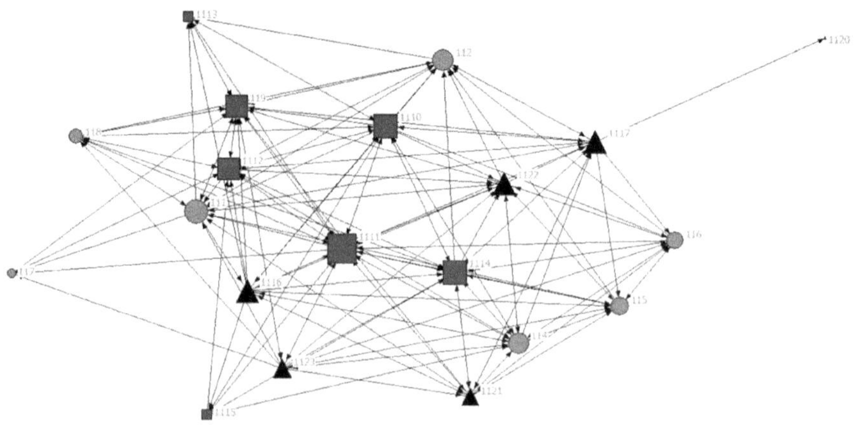

FIGURE 3.2 Collaboration network among participants in the Comitê Médio Paraíba water council.

We can infer several characteristics of the network from observing its graph. First, we observe that there are no disconnected participants in the basin council, that is, that all participants maintain relations with at least some others. Only one actor (1120) has a single strong link coming from another actor in its own sector. The fact that most nodes are connected to several others was to be expected considering two factors: (i) the method we used to construct the network was based on presenting actors with the list of participants in the basin, and although they could pinpoint actors that they didn't know, this method makes it unlikely that totally disconnected participants will be observed; (ii) the network studied is quite small and tied to a forum that recurrently meets, which makes it likely that participants would know most of the other participants.

Additionally, we observe that some actors seem to receive more incoming ties than others, that is, their shape is bigger than that of others. However, it also seems that actors from different sectors are among the biggest shapes which suggests that centrality is well shared across sectors.

Network visualization gives a quick overview of the general configuration of the network but it should not be taken as demonstrating any particular hypothesis. It is important to keep in mind that different algorithms can present the network visualization in significantly different ways and thus must be used with caution. By looking at a network graph, it is not possible to provide specific information on the inner configurations of the network, to investigate those it is necessary to turn to ERGM.

ERGM results

In this study we used the freely accessible software MPNet for our ERGM.[1] As we have seen above, the selection of attributes and configurations was guided by our theoretical framework and research questions. We focused on two configurations, reciprocity and AinS (Table 3.2). We also included four attributes to explain centrality (involvement, time, forums, and influence) and one attribute to assess homophily (sector). The results are summarized in Table 3.2 below.

TABLE 3.2 Exponential Random Graph Model results

	Parameter	*Standard error*
ArcA	−5.0917*	0.952
Reciprocity	1.7591*	0.327
AinS	1.4586*	0.527
Involvement	0.1780	0.139
Time	0.0870	0.068
Forums	0.0915	0.085
Influence	0.0000	0.000
Sector homophily	0.1937	0.208

* indicates significance.

After the model is obtained, 'goodness-of-fit' (GOF) tests need to be run to test how well the model reproduces the actual network of empirical data (see Box 3.1). Our model presented a reasonably good fit and was stable. However, the GOF test indicated that we were missing clustering effects. We acknowledge that the network surely contains clusters but the GOF indicates that we captured the effects that we were interested in, in line with our theoretical framework.

What does it mean in terms of participation?

In analyzing the results of the models, it is important to keep in mind that what doesn't appear to be significant can be as important as what appears to be significant. Both provide us with insights on our hypotheses. Through this ERGM we investigated two main issues: whether certain configurations were significant in explaining the structure of our network and whether different attributes could help us explain why actors linked up to certain others. As we observe from the results, none of the selected attributes help us explain why certain actors are more sought-after than others. In other words, none of these factors are predictors of incoming ties for a given actor. By contrast, the two configurations selected, reciprocity and AinS, are significant in explaining the general configuration of the network.

Involvement appears not to be significant, and therefore hypothesis 1 should be rejected. This result must be interpreted with a caveat: involvement was self-reported and actors might have overestimated their actual level of involvement, or such level of involvement might not be perceived as such by others. Among the participants less likely to be deeply involved in the council's daily work we find municipalities (Mancilla García et al., 2019). Yet participants in the council are likely to try and collaborate with municipalities regardless of their level of involvement since their role is fundamental in organizing outreach activities to raise awareness on water related issues, which hints at one reason why involvement does not appear as a significant attribute to explain centrality in the collaboration network.

Time spent over the years as a participant in the council does not help explain why some actors are more sought-after than others when it comes to collaboration, which leads us to reject hypothesis 2. This suggests that the council remains open to newcomers and that participants don't rely on specific others to make decisions. This also suggests that the council does not seem prone to elite capture, a situation when a group of participants establishes itself as dominant in a given forum and maintains that position through time (Fritzen, 2007; Labonte, 2012; Saito-Jensen et al., 2010; Wong, 2017). This should be understood in light of the specificities of the council studied. Indeed, the council is a small forum that covers a relatively small region where water stakeholders tend to know each other (regardless of whether they participate in the council or not). Moreover, the council focuses its activities around environmental education and communication and provides support for small municipalities to manage waste waters.

These are issues that tend to drive consensus and thus help to explain why we don't observe elite capture phenomena.

The fact that the number of forums attended is not significant in explaining why actors link to others is surprising since it contrasts with previous results on the dynamics at play in other councils in the area (Mancilla García and Bodin, 2018). The number of forums attended might serve to explain influence, or information degree centrality, but it does not appear as relevant to explain collaboration. This result must also be interpreted in light of the specifics of the council studied, which, as we said before is a small council focused essentially on activities developed in its area of jurisdiction. This means that linking to other forums – such as other councils or the state committee for water resources – might appear as having less impact on the decisions made in this forum, which functions relatively independently. Indeed, while the decisions made in higher up forums can potentially affect the Council of Médio Paraíba, the nature of the daily activities in which it engages might make such effect negligible for participants. Hypothesis 3 is thus rejected.

The last of the attributes for which we tested whether it had an effect on explaining centrality is influence, which we also found not to be significant, rejecting hypothesis 4. This suggests that participants don't seek to align their positions with those participants perceived as the most influential. This might be because participants prefer to work with those who share their vision or ideas – regardless of their level of influence – or because the nature of the decisions made is perceived as not necessarily requiring influential actors to be implemented.

We also tested whether homophily – when it comes to sectors – was a significant factor. The fact that it isn't means that people don't link up exclusively with others in their sector, and thus hypothesis 6 is also rejected. This is an important finding when it comes to evaluating how participatory models function since it suggests that people go beyond their original group to collaborate. In our particular case, it means that people from different sectors align with others in different sectors, not only with those in their own sector. This means that the council seems to effectively encourage interaction among participants from different sectors.

As we saw in the theoretical section, the attributes that we selected are typically studied in the literature to try and explain 'popularity.' That said, we could have tested for other attributes such as whether certain participants are perceived as knowledgeable and trustworthy as other studies have explored (Leach et al., 2014). We acknowledge that this is a limitation of our study.

What appears as significant in our results are the two configurations tested, namely, AinS and reciprocity. The significance of AinS indicates that the more central an actor is, the more ties it tends to receive, confirming hypothesis 5. Since none of the selected attributes explain why actors are central, this suggests that being central makes an actor more sought after. We also observe that reciprocity is significant which means that actors tend to align with those who align with them, that is, that participants tend to decide together with those they are aligning with, rather than follow a leader that ignores them. This suggests that horizontal structures are, at least to a certain extent, important in the council and confirms hypothesis 7.

Conclusion

The results of our study indicate that the Médio Paraíba water council is indeed a network and not a random collection of actors. The prominent processes in this network are basic social processes such as reciprocation – that is, the process of alignment tends to be mutual – and popularity – those who are already popular tend to be more popular. This also seems to confirm our observation of the dynamics at play in the council, where participants know each other and some of them seemed to be well-known in the water governance sector of the region. Additionally, the fact that the attributes tested were not significant suggests the network presents the characteristics of a horizontal deliberation network. In such a setting, actors do not pay much attention to the interests and stakes others are associated with when they engage in network building. A reason for this might be that actors are able to look beyond these potential barriers to cooperation when they get together and discuss common issues. Another reason could be that they do not perceive this forum as dealing with controversial issues, but rather focusing on issues where consensus is within reach, and/or on issues not perceived as involving high stakes.

By providing the methodological tool to assess how actors are connected and, for example, whether some actors are better connected than others, SNA constitutes a useful approach to evaluate whether participatory forums manage to fulfill their objectives of inclusion and horizontal deliberation. Additionally, the use of statistical network models such as ERGMs allows for the testing of diverse sets of hypotheses including not only the effect of actors' attributes in the formation of social ties, but also the role of specific social network configurations – such as centrality – to explain the way actors relate to each other.

SNA is an appropriate tool for natural resource governance scholars more generally, that is, beyond focusing on participation, since it can provide a nuanced picture of the relationships at play in a given governance setting. It can reveal certain factors that make some actors more influential than others and the characteristics of marginalized or excluded actors. Although beyond the scope of this chapter, researchers can use SNA to investigate the subtleties of relationships by focusing on networks in which there are at the same time relations of conflict and collaboration (see, e.g., Robins et al., 2011). If it is possible to obtain longitudinal network data, researchers can additionally study how relationships evolve with time, thereby providing the opportunity to disentangle causes and effects in social tie formation processes and in associated changes of certain actors' attributes such as their values and preferences. This enables understanding of the kind of cooperative and competitive processes that predominate in a given governance setting, why and how diverse actors find ways of inclusion, and what kind of social relational patterns lead to more inclusive and sustainable outcomes.

Acknowledgments

This project received funding from the European Union's Horizon 2020 research and innovation programme under the Marie Skłodowska-Curie grant agreement 660394. We would like to thank Andrea Nascetti for his help with maps and to all participants for taking part in this project. We extend our thanks to Paúl Cisneros and Matthew Hamilton for their comments on the surveys.

Note

1 The software and manual can be found at www.melnet.org.au/pnet/

References

Akbulut, B. and Soylu, C. (2012) 'An inquiry into power and participatory natural resource management', *Cambridge Journal of Economics*, vol. 36, no. 5, pp. 1143–1162.

Alcañiz, I. and Berardo, R. (2016) 'A network analysis of transboundary water cooperation in La Plata Basin', *Water Policy*, vol. 18, no. 5, pp. 1120–1138.

Alexander, S. M. and Armitage, D. (2015) 'A social relational network perspective for MPA science', *Conservation Letters*, vol. 8, no. 1, pp. 1–13.

Altschuler, D. and Corrales, J. (2012) 'The spillover effects of participatory governance: Evidence from community-managed schools in Honduras and Guatemala', *Comparative Political Studies*, vol. 45, no. 5, pp. 636–666.

Angst, M., Widmer, A., Fischer, M. and Ingold, K. (2018) 'Connectors and coordinators in natural resource governance: Insights from Swiss water supply', *Ecology and Society*, vol. 23, no. 2, p. 1

Barnaud, C. and Van Paassen, A. (2013) 'Equity, power games and legitimacy: Dilemmas of participatory natural resource management', *Ecology and Society*, vol. 18, no. 2: 21, http://dx.doi.org/10.5751/ES-05459-180221.

Barnes, M. L., Lynham, J., Kalberg, K. and Leung, P. (2016) 'Social networks and environmental outcomes', *Proceedings of the National Academy of Sciences*, vol. 113, no. 23, pp. 6466–6471.

Basurto, X., Blanco, E., Nenadovic, M. and Vollan, B. (2016) 'Integrating simultaneous prosocial and antisocial behavior into theories of collective action', *Science Advances*, vol. 2, no. 3, p. e1501220.

Beierle, T. C. and Cayford, J. (2002) *Democracy in Practice: Public Participation in Environmental Decisions*. Resources for the Future, Washington, DC.

Berardo, R. and Scholz, T. J. (2010) 'Self-organizing policy networks: Risk, partner selection, and cooperation in estuaries', *American Journal of Political Science*, vol. 54, no. 3, pp. 632–649.

Bodin, Ö. (2017) 'Collaborative environmental governance: Achieving collective action in social-ecological systems', *Science*, vol. 357, no. 6352, p. eaan1114.

Bodin, Ö. and Prell, C. (2011) *Social Networks and Natural Resource Management: Uncovering the Social Fabric of Environmental Governance*. Cambridge University Press, Cambridge.

Bulkeley, H. and Mol, J. P. A. (2014) 'Participation and environmental governance: Consensus, ambivalence and debate', *Environmental Values*, vol. 12, no. 2, pp. 143–154.

Cornwall, A. and Coelho, S. V. (2006) *Spaces for Change? The Politics of Citizen Participation in New Democratic Arenas*. Zed Books, London.

Dahl, R. (1994) 'A democratic dilemma: System effectiveness vs. citizen participation', *Political Science Quarterly*, vol. 109, no. 1, pp. 23–34.

Davies, A. L. and White, M. R. (2012) 'Collaboration in natural resource governance: Reconciling stakeholder expectations in deer management in Scotland', *Journal of Environmental Management*, vol. 112, pp. 160–169.

Diduck, A. and Sinclair, A. J. (2002) 'Public involvement in environmental assessment: The case of the nonparticipant', *Environmental Management*, vol. 29, no. 4, pp. 578–588.

Edelenbos, J., van Buuren, A. and van Schie, N. (2011) 'Co-producing knowledge: Joint knowledge production between experts, bureaucrats and stakeholders in Dutch water management projects', *Environmental Science and Policy*, vol. 14, no. 6, pp. 675–684.

Fischer, M. and Sciarini, P. (2016) 'Drivers of collaboration in political decision making: A cross-sector perspective', *The Journal of Politics*, vol. 78, no. 1, pp. 63–74.

Freeman, L. C. (2004) *The Development of Social Network Analysis – A Study in the Sociology of Science*. Empirical Press, Vancouver.

Fritzen, S. A. (2007) 'Can the design of community-driven development reduce the risk of elite capture? Evidence from Indonesia', *World Development*, vol. 35, no. 8, pp. 1359–1375.

Furlong, S. R. (1997) 'Interest group influence on rule making', *Administration & Society*, vol. 29, no. 3, pp. 325–347.

Henry, A. D. (2011) 'Ideology, power, and the structure of policy networks', *Policy Studies Journal*, vol. 39, no. 3, pp. 361–383.

Henry, A. D., Lubell, M. and McCoy, M. (2012) 'Survey-based measurement of public management and policy networks', *Journal of Policy Analysis and Management*, vol. 31, no. 2, pp. 432–452.

Ingold, K. and Fischer, M. (2014) 'Drivers of collaboration to mitigate climate change: An illustration of Swiss climate policy over 15 years', *Global Environmental Change*, vol. 24, no. 1, pp. 88–98.

Ingold, K. and Leifeld, P. (2016) 'Structural and institutional determinants of influence reputation: A comparison of collaborative and adversarial policy networks in decision making and implementation', *Journal of Public Administration Research and Theory*, vol. 26, no. 1, pp. 1–18.

Irvin, R. and Stansbury, J. (2004) 'Citizen participation in decision making: Is it worth the effort?', *Public Administration Review*, vol. 64, no. 1, pp. 55–65.

Jacobs, K., Lebel, L., Buizer, J., Addams, L., Matson, P., McCullough, E., Garden, P., Saliba, G. and Finan, T. (2010) 'Linking knowledge with action in the pursuit of sustainable water-resources management', *Proceedings of the National Academy of Sciences of the United States of America*, vol. 113, no. 17, pp. 4591–4596.

König, T. and Bräuninger, T. (1998) 'The formation of policy networks: Preferences, institutions and actors' choice of information and exchange relations', *Journal of Theoretical Politics*, vol. 10, no. 4, pp. 445–471.

Labonte, M. T. (2012) 'From patronage to peacebuilding? Elite capture and governance from below in Sierra Leone', *African Affairs*, vol. 111, no. 442, pp. 90–115.

Leach, W. D., Weible, M. C., Vince, R. S., Siddiki, N. S. and Calanni, J. C. (2014) 'Fostering learning through collaboration: Knowledge acquisition and belief change in marine aquaculture partnerships', *Journal of Public Administration Research and Theory*, vol. 24, no. 3, pp. 591–622.

Leifeld, P. (2013) 'Reconceptualizing major policy change in the advocacy coalition framework: A discourse network analysis of German pension politics', *The Policy Studies Journal*, vol. 41, no. 1, pp. 169–198.

Lubell, M. (2007) 'Familiarity breeds trust: Collective action in a policy domain', *Journal of Politics*, vol. 69, no. 1, pp. 237–250.

Lubell, M., Henry, D. A. and McCoy, M. (2010) 'Collaborative institutions in an ecology of games', *American Journal of Political Science*, vol. 54, no. 2, pp. 287–300.

Lubell, M., Mewhirter, M. J., Berardo, R. and Scholz, T. J. (2017) 'Transaction costs and the perceived effectiveness of complex institutional systems', *Public Administration Review*, vol. 77, no. 5, pp. 668–680.

Lubell, M., Scholz, J., Berardo, R. and Robins, L. G. (2012) 'Testing policy theory with statistical models', *Policy Studies Journal*, vol. 40, no. 3, pp. 351–374.

Lund, J. F. and Saito-Jensen, M. (2013) 'Revisiting the issue of elite capture of participatory initiatives', *World Development*, vol. 46, pp. 104–112.

Lusher, D., Koskinen, H. J. and Robins, G. (2013) *Exponential Random Graph Models for Social Networks: Theory, Methods, and Applications*. Cambridge University Press, Cambridge.

Mancilla García, M. and Bodin, Ö. (2018) 'Participation in multiple decision making water governance forums in Brazil enhances actors' perceived level of influence', *Policy Studies Journal*, vol. 47, no. 1, pp. 27–51.

Mancilla García, M. and Bodin, O. (2019) 'Participatory water basin councils in Peru and Brazil: Expert discourses as means and barriers to inclusion', *Global Environmental Change*, vol. 55, pp. 139–148.

Mancilla García, M., Hileman, J., Bodin, Ö., Nilsson, A. and Jacobi, R. P. (2019) 'The unique role of municipalities in integrated watershed governance arrangements : A new research frontier', *Ecology & Society*, vol. 24, no. 1: 28. https://doi.org/10.5751/ES-10793-240128.

McAllister, R. R. J., Robinson,J. C., Brown, A., Maclean, K., Perry, S. and Liu, S. (2017) 'Balancing collaboration with coordination: Contesting eradication in the Australian plant pest and disease biosecurity system', *International Journal of the Commons*, vol. 11, no. 1, pp. 330–354.

Milo, R., Shen-Orr, S., Itzkovitz, S., Kashtan, N., Chklovskii, D. and Alon, U. (2002) 'Network motifs: Simple building blocks of complex networks', *Science*, vol. 298, no. 5594, pp. 824–827.

Newig, J., Challies, E., Jager,W. N., Kochskaemper, E. and Adzersen, A. (2018) 'The environmental performance of participatory and collaborative governance: A framework of causal mechanisms', *Policy Studies Journal*, vol. 46, no. 2, pp. 269–297.

Newig, J. and Kvarda, E. (2012) 'Participation in environmental governance: Legitimate and effective?' in K. Hogl, E. Kvarda, R. Nordbeck, and M. Pregernig (eds) *Environmental Governance. The Challenge of Legitimacy and Effectiveness*. Edward Elgar, Cheltenham.

Nowell, B. L., Velez, K. A., Hano, C. M., Sudweeks, J., Albrecht, K. and Steelman, T. (2018) 'Studying networks in complex problem domains: Advancing methods in boundary specification', *Perspectives on Public Management and Governance*, vol. 1, no. 4, pp. 273–282.

Ostrom, E. (1990) *Governing the Commons: The Evolution of Institutions for Collective Action*. Cambridge University Press, Cambridge.

Robins, G. (2015) *Doing Social Network Research: Network-Based Research Design for Social Scientists*. SAGE, London.

Robins, G., Bates, L. and Pattison, P. (2011) 'Network governance and environmental management: Conflict and cooperation', *Public Administration*, vol. 89, no. 4, pp. 1293–1313.

Robins, G., Lewis, M. J. and Wang, P. (2012) 'Statistical network analysis for analyzing policy networks', *Policy Studies Journal*, vol. 40, no. 3, pp. 375–401.

Sabatier, P. (1988) 'An advocacy coalition framework of policy change and the role of policy-oriented learning therein', *Policy Sciences*, vol. 21, no. 2–3, pp. 129–168.

Sabatier, P. and Jenkins-Smith, H. (1999) 'The advocacy coalition framework: An assessment' in P. Sabatier (ed.) *Theories of the Policy Process*. Westview Press, Boulder, CO.

Saito-Jensen, M., Nathan, I. and Treue, T. (2010) 'Beyond elite capture? Community-based natural resource management and power in Mohammed Nagar village, Andhra Pradesh, India', *Environmental Conservation*, vol. 37, no. 3, pp. 327–335.

Sandström, A., Crona, B. and Bodin, Ö. (2014) 'Legitimacy in co-management: The impact of preexisting structures, social networks and governance strategies', *Environmental Policy and Governance*, vol. 24, no. 1, pp. 60–76.

Scott, T. A. and Ulibarri, N. (2019) 'Taking network analysis seriously: Methodological improvements for governance network scholarship', *Perspectives on Public Management and Governance*, vol. 2, no. 2, pp. 89–101.

Singleton, S. (2000) 'Co-operation or capture? The paradox of co-management and community participation in natural resource management and environmental policy-making', *Environmental Politics*, vol. 9, no. 2, pp. 1–21.

Weible, C. M. (2005) 'Beliefs and perceived influence in a natural resource conflict: An advocacy coalition approach to policy networks', *Political Research Quarterly*, vol. 58, no. 3, pp. 461–475.

Weible, C. M. and Sabatier, A. P. (2009) 'Coalitions, science, and belief change: Comparing adversarial and collaborative policy subsystems', *Policy Studies Journal*, vol. 37, no. 2, pp. 195–212.

Wong, S. (2017) 'Challenges to the elite exclusion – inclusion dichotomy — Reconsidering elite capture in community-based natural resource management', *South African Journal of International Affairs*, vol. 20, no. 3, pp. 379–391.

Yackee, J. W. and Yackee, W. S. (2006) 'A bias towards business? Assessing interest group influence on the U.S. bureaucracy', *The Journal of Politics*, vol. 68, no. 1, pp. 28–139.

Young, J. C., Jordan, A., Searle, R. K., Butler, A., Chapman, S. D., Simmons, P. and Watt, D. A. (2013) 'Does stakeholder involvement really benefit biodiversity conservation?', *Biological Conservation*, vol. 158, pp. 359–370.

4

ANALYZING NATURAL RESOURCE GOVERNANCE WITH THE SOCIAL-ECOLOGICAL SYSTEMS FRAMEWORK

Stefan Partelow

Introduction

Analyzing natural resource governance, as popularized by the political economist Elinor Ostrom, includes systems thinking, understanding what influences human behavior and examining the links between social and environmental change. However, Ostrom did not start there. Rather, beginning in the 1980s (Anderies and Janssen, 2012), the core questions were, 'Can people develop their own cooperative solutions to use shared resources?' 'How and why are some groups able to do this, and others not?'

So, why are these interesting questions to ask? At the time, economic rational choice theory hypothesized that individuals would not be motivated to reduce resource use in the short term, only to let others gain from harvesting. Scholarship by early 20th century economists such as Warming (1911), Gordon (1954), Scott (1955) and Hardin (1968), an ecologist, forewarned that the use of shared resources was a problem humans were inherently disinclined to solve. Cooperative strategies were presumed to be irrational from an individual perspective. Self-interest would bring individual gains, and it was not considered that individuals would invest in gains to be shared across a group, especially when it is unclear if individuals are free-riding on the work of others.

Sustainability is an important concept for evaluating the consistent and long-term availability of natural resource use under assumptions of economic rational choice theory. Early economists were some of the first to theoretically address concerns for the sustainability of natural resource use, tracing back to forestry concerns in Germany in 1700s (Caradonna, 2014). For example, in fisheries, all fishers want stable and long-term fish populations, but the interest of individuals was assumed to be maximization in the short term. If individual fishers do this, it undermines the goal of all fishers.

In common-pool resource systems such as fisheries, ground water and forestry, rational economic logic would presume that individual resource use behavior will lead to unsustainable patterns over time, depleting a resource. Solving these problems requires governance, and governance is a concept societies have developed to guide human behavior towards desired outcomes (Ostrom, 1998; Stoker, 1998). Although cooperative behavior has been a key survival strategy for humans throughout our history, many aspects of human behavior are not programmed to be cooperative. We have many inherent perceptual and decision-making biases that favor individual gains, often at the expense of others. However, when such behavior can be incentivized to align with group goals through governance, through the appropriate combinations of rules and norms, progress towards sustainable outcomes is more likely.

The concept of governance, what it is and how it works, can be viewed in many ways (Kooiman et al., 2008; Ostrom, 1990; Stoker, 1998; van Assche et al., 2014). From Ostrom's perspective, and within the fields of political science and institutional economics, governance is a set of interacting institutions to guide behavior towards group goals. Institutions in the social sciences do not refer to organizations, they refer to sets of rules and norms (Klein, 2000). In turn, governance does not mean government. Governments, in the sense of Western state bureaucracies, can influence institutions for governance, but they are certainly not the only influence on governance or the broader institutions that 'govern' or influence human behavior.

With hindsight on more than a half century of commons scholarship, it is now easier to view the pessimism of economic rational choice theory as rather simplistic and incomplete if it is considered the only logic guiding human behavior that influences resource use. In the mid-20th century, knowledge in the field of institutional and behavioral economics was limited. Nonetheless, the reality of global sustainability challenges is real, and many natural resources are overexploited (Steffen et al., 2015). Existing theories of resource use behavior have helped provide a baseline for hypothesis testing and further understanding of what drives sustainability challenges, but further research is still needed to more effectively understand what institutional arrangements can best 'govern' behavior towards sustainability while avoiding tendencies to implement panacea approaches. It is important to note that the logic of rational choice theories, and the portrayal of *Homo economicus* (Thaler, 2000), do not make a completely incorrect assumption about what happens if human behavior is left unguided by rules and norms. Rather, this is only one piece of the puzzle trying to understand what the mechanisms driving human behavior are. Why do resource users act in the interest of the group under some combinations of rules and norms, and in other cases they act individualistically? Are the successful combinations of rules and norms transferable to other cases?

Much of the motivation for commons scholarship, since the 1980s, has stemmed from Hardin's (1968) conclusions in 'The Tragedy of the Commons.' As noted above, he imagined a pasture open access to a group of economically rational resource users, where only famine and disease had historically kept self-interested resource use restrained in the face of rivalry and excludability challenges. Hardin suggested that there were only two viable governance solutions to ensure natural resource sustainability: (1) well-enforced top-down state control that coerces sustainable use behavior or (2) privatization. The logic of privatization is that if property rights (e.g., use or access rights) are given to individuals, they will be incentivized to use the resources efficiently, in part, by internalizing what classical economists call 'externalities' of the group back to the individual. Of course, we now know that 'externalities' are not external costs, they are actual costs that are spread to the group (e.g., pollution) that have historically not been included as production costs in classic economic models. Moreover, the logic of privatization mitigates the worry about rivalry or freeriding, better fostering resource stewardship when investment efforts for sustainable use are protected from others.

Aiming to test Hardin's hypotheses, commons scholarship has examined what governance arrangements (sets of formal (rules) and informal (norms) institutions) can lead to sustainable and desired outcomes. Commons scholars also identified another governance model missed by Hardin. Common property arrangements are private ownership at the group or community level, and are often self-organized from the bottom-up through collective action (CA) (Agrawal, 2001). Commons research has also examined under what social and environmental conditions it is more or less likely for successful governance solutions to emerge overtime (simplistically as combinations of common property, privatization, or state control), or to be developed in collaboration with local communities and external organizations (e.g., governments, nongovernmental organizations [NGOs]). Many of the findings in commons scholarship sit in stark contrast to Hardin, showing that people can and do self organize cooperative solutions, often in the form of institutions with common property arrangements to use shared resources in many diverse contexts.

Although numerous public policy efforts using state enforcement or privatization have led to desired outcomes, many have not. Many of such policies do not work or fit the context to which they are applied. Two important lessons have been learned from commons scholarship and Ostrom's research leading to the social-ecological systems framework (SESF). First, generic policy solutions do not fit all cases. This is due to the social and ecological diversity of natural resources scenarios that exist, leading to highly heterogeneous impacts and outcomes from generic prescriptions (Ostrom, 2007). Second, many of the public policies fail to recognize the role of CA among local stakeholders and communities. This could be in the form of replacing or negating preexisting community-based governance institutions when implementing state control or privatizations, and/or in neglecting the importance of incorporating local stakeholders and communities into decision-making processes, by better understanding how and why people can work together more effectively.

Collective action theory contributes knowledge on both of the points above, either (1) that communities are capable of self-organizing institutions for governance themselves (under certain conditions, as explored below) or (2) that in cases where privatizations or state interventions are a necessary part of the governance system, that understanding how stakeholders can work together and be included in decision-making processes is critical for developing collaborative governance that works in practice, such as in co-management.

Among many contributions, collective action theory has helped reveal the flaws in Hardin's assumptions. First, although rivalry and non-excludability are inherent features, Ostrom showed that resource users can avoid overuse by self-organizing diverse sets of institutions for governance. Second, she showed that resource use systems have many more influential variables than Hardin assumed (both social and ecological) (Ostrom, 2007). Third, Ostrom showed that, in many cases, cooperative decisions were made by resource users instead of self-interested economic ones due to clever institutional arrangements. This included identifying that locally adapted institutions, sets of rules and norms, evolve and change overtime to guide behavior in a way that fits the context. The process of adaptive institutional change, although imperfect and often contentious, is an important part of natural resource governance.

The central hypothesis or premise of collective action theory is that people who share commons resources, either public goods (e.g., shared infrastructure, climate) or common-pool resources (e.g., fish, trees, water), are able to work together and cooperate to achieve group goals when institutional arrangements are adapted over time to fit the context. However, this can be easier or more difficult depending on the social and environmental context. A central task of commons scholarship, leading to the SESF, has been to identify the variables that influence self-organization and CA, and under what conditions each variable makes it easier or more difficult.

Understanding natural resource governance, as mentioned in the beginning of this chapter, involves understanding how many different variables interact within a system. It is unlikely that any single governance theory or framework will capture the full spectrum of system interactions and influences that guide human behavior or all the linkages between social and environmental change. Collective action theory is one useful tool for solving the governance puzzle, but many others exist and are needed to provide analytical diversity. The reality of natural resource governance in many parts of the world, is a mixture of state oversight, private property rights and CA among local stakeholders. All these aspects require collaboration, and although collective action theory is only one tool for understanding governance, it has gained increasing prominence as many state governments are reverting away from exclusive top-down control or property rights allocations, and moving towards collaborative governance (Bodin, 2017), most commonly seen as various forms of co-management (Berkes, 2009).

Both the design principles (Ostrom, 1990) (featured in Chapter 1) and subsequent SESF (McGinnis and Ostrom, 2014; Ostrom, 2007, 2009; Partelow, 2018) are attempts to identify the variables that influence CA, but they also provide insights into what influences collaborative governance arrangements such as co-management. Ostrom's design principles make explicit theoretical claims on how each variable is influential, although they only include social variables. The SESF presents a similar list of variables, including ecological variables, presented conceptually as an interacting system. However, the SESF does not include explicit claims on how each variable influences CA. Thus, although it is structured and developed as a theoretical framework to examine community-based CA processes, it is often claimed to be non-theoretical, to both broaden its usefulness and for building new theories with the variables (see discussion in Partelow, 2018).

The SESF

The SESF serves numerous academic purposes. One contribution is bringing a systems thinking perspective to natural resource governance, visualized in Figure 4.1. Although the framework is grounded in collective action theory, its name and visual presentation of first-tier variable interactions (Figure 4.1) have helped catalyze its notoriety and contributions to discourse and theory on social-ecological systems. In addition, the SESF is an academic tool designed to help solve the data comparability problem in the environmental social sciences. There is a need to robustly compare case study research in order to generalize findings to inform natural resource governance theories. Comparing independent and dependent variables across diverse cases requires common variables to compare. Focused case study research often provides rich and detailed data on different contexts, but often measures variables unique to the case. On the other hand, it is hard to decide what the common variables and measurement

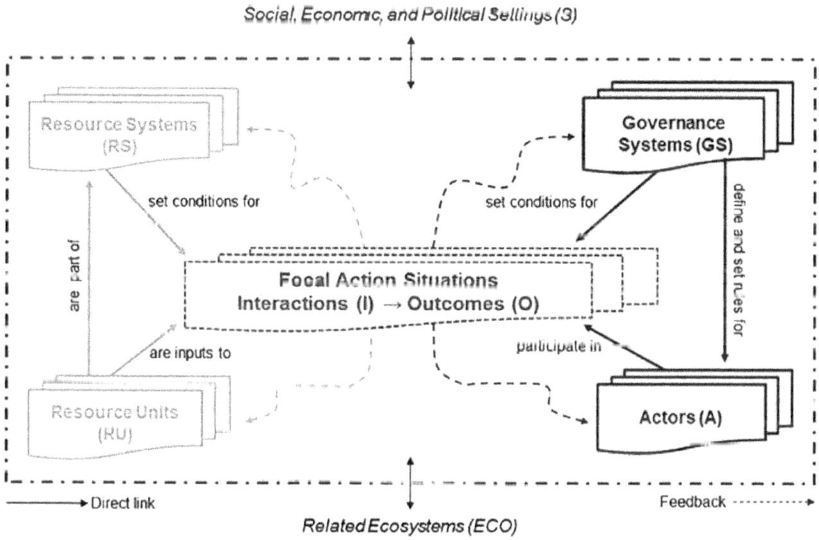

FIGURE 4.1 The social-ecological systems framework ('SESF'). The first-tier variables and visual conceptualization of the SESF from McGinnis and Ostrom (2014).

methodologies should be used, in a way that does not limit the depth of case study research.

The SESF attempts to provide a non-context specific list of common variables, as general and adaptable as possible for specific cases, to facilitate comparability between cases that use the variables (McGinnis and Ostrom, 2014; Ostrom, 2009). In addition, the framework is designed to be used diagnostically, similar to the logic of how doctors diagnose patients. The goal is to identify a natural resource use problem and then systematically examine the variables potentially influencing the problem. A diagnostic approach assumes there are undesired outcomes, either social (e.g., inequities, low income, poverty traps) or ecological (e.g., resource decline, pollution). This would allow the analyst (e.g., government officials, local leaders, scholars) to give the appropriate treatment (e.g., formal or informal institutional developments and change). Thus, the SESF can be viewed as a checklist, where each variable is checked systematically with appropriate methods, one by one, to examine their influence, identify problems, and then assess institutional solutions in response.

In reality, using the SESF as a diagnostic tool, and solving the comparability problem, is only a starting point. Common variables provide the means to do this, what is often referred to as a 'common language' to communicate findings between different fields and disciplines, but this is not the only problem that needs solving. Two reviews of the SESF literature show that many studies use the framework, but in methodologically heterogeneous ways (Partelow, 2018; Thiel et al., 2015). This suggests that transparent communication of variable definitions and methodologies is a critical gap for ensuring comparable diagnostic research, for both data collection and analysis. This multistep process of applying the SESF can be referred to as 'operationalizing' the framework (Leslie et al., 2015; Partelow et al., 2018b). This includes developing the research questions, defining the social and ecological boundaries of the study, defining the variables to be measured and then developing empirical data collection and analysis methodologies.

The reality is that 'there are no general methods, guidelines, or procedures for applying the SESF, although numerous articles have provided conceptual guidance' (Partelow, 2018: 11). This is both a benefit and a shortcoming. As a benefit, the framework is very adaptable, and the framework has been applied in many diverse contexts and in new ways useful to the literature. As a downside, the literature applying the framework is methodologically heterogeneous, largely non-comparable, and there is confusion over how it can be applied as well as its claim to be a SESF despite its clear origin and evolution as a theoretical framework for collective action theory. There is no correct way to apply it, and the literature reviews show that it can be utilized in many ways to answer a diverse array of research questions. Methodological innovations are encouraged with the suggestion that transparency in communicating how and why the framework is applied is a priority (e.g., contributes to what theory or empirical goal). For example, many studies apply the framework to examine collective action theory, but many other papers apply the framework to characterize systems as social-ecological systems without a theoretical basis or under a broader notion of social-ecological systems theory. These papers make different types of contributions to the literature and development trajectories for the SESF.

The SESF is a set of nested first-tier (Figure 4.1) and second-tier variables (Table 4.1). The design of nested tiers was thought to be malleable, with the

TABLE 4.1 First and second-tier variables of the social-ecological systems framework (SESF) from McGinnis and Ostrom (2014)

Social, Economic, and Political Settings (S)	
S1 – Economic development. S2 – Demographic trends. S3 – Political stability. S4 – Other governance systems. S5 – Markets. S6 – Media organizations. S7 – Technology.	
Resource Systems (RS)	**Governance Systems (GS)**
RS1 – Sector (e.g., water, forests, pasture)	GS1 – Government organizations
RS2 – Clarity of system boundaries	GS2 – Nongovernment organizations (NGOs)
RS3 – Size of resource system	GS3 – Network structure
RS4 – Human-constructed facilities	GS4 – Property-rights systems
RS5 – Productivity of system	GS5 – Operational-choice rules
RS6 – Equilibrium properties	GS6 – Collective-choice rules
RS7 – Predictability of system dynamics	GS7 – Constitutional-choice rules
RS8 – Storage characteristics	GS8 – Monitoring and sanctioning rules
RS9 – Location	
Resource Units (RU)	**Actors (A)**
RU1 – Resource unit mobility	A1 – Number of relevant actors
RU2 – Growth or replacement rate	A2 – Socioeconomic attributes
RU3 – Interaction among resource units	A3 – History or past experiences
RU4 – Economic value	A4 – Location
RU5 – Number of units	A5 – Leadership/entrepreneurship
RU6 – Distinctive characteristics	A6 – Norms (trust-reciprocity)/social capital
RU7 – Spatial and temporal distribution	A7 – Knowledge of SES/mental models
	A8 – Importance of resource (dependence)
	A9 – Technologies available
Interactions (I)	**Outcomes (O)**
I1 – Harvesting	O1 – Social performance measures
I2 – Information sharing	O2 – Ecological performance measures
I3 – Deliberation processes	O3 – Externalities to other SESs
I4 – Conflicts	
I5 – Investment activities	
I6 – Lobbying activities	
I7 – Self-organizing activities	
I8 – Networking activities	
I9 – Monitoring activities	
I10 – Evaluative activities	
Related Ecosystems (ECO)	
ECO1 – Climate patterns ECO2 – Pollution patterns ECO3 – Flows into and out of SES	

possibility of adding new variables within existing tiers or to new third or fourth-tiers where new empirical work supports it (Ostrom, 2007, 2009). Guidance for adding new variables has been explored considering the framework's structural logic or ontology (see page 14 in Frey and Cox, 2015), as well as what may need to be considered as inclusion criteria (e.g., relation to collective action theory) for adding new variables (see Partelow, 2018). Each second-tier variable of the framework has been included historically because it has been empirically demonstrated to influence CA processes in some way.

Applying the SESF: crab fishing, mangroves, and co-management in Bragança, Brazil

In this section, the results of research applying the SESF to a mangrove crab fishery in Brazil are presented and discussed. A more detailed data collection and analysis methodology as well as an extended background to the case study can be found specifically in Partelow et al. (2018a) and generally in Saint-Paul and Schneider (2010). The purpose of this case study section is to show how the SESF can be used as a diagnostic tool to examine variables influencing the governance of the Brazilian Marine Extractive Reserve (RESEX) co-management program in Bragança. Methodologically, this case study used the SESF to develop qualitative data collection questions for interviews with the case's stakeholders, and then as a qualitative data analysis tool to help code the content of the interviews for data organization and analysis.

RESEX co-management requires stakeholders at multiple social organizational levels to work together and develop institutions for governance through co-management. RESEX is a broad legal framework with the aim of jointly advancing ecological conservation through social empowerment within local natural resource dependent communities in rural Brazil. This case study looks at the evolution of how numerous variables have interacted over time to shape the challenges facing RESEX implementation, much of which is leveraged on effective cooperation between different stakeholder groups, and particularly on CA within local communities who need to self-organize their representation and goals for co-management within a committee tasked with decision-making. CA theory is a useful lens to view the social and ecological variables that are influencing cooperative behavior, and thus overall outcomes, within the RESEX program.

The Caeté-Taperaçu (CT) Marine RESEX

On the coast of northeast Brazil, the rural city of Bragança is situated within the second largest continuous mangrove forest in the world; one of many rural areas in the region with high dependencies on local natural resources, home to more than 100,000 people. Mangrove resources in the area are a primary source of food and livelihoods for more than 40,000 rural inhabitants who generally lack access to alternative education and economic opportunities. Crab fishing is the most important natural resource, but over 20 important resources or uses of the mangrove forests have been documented (Glaser, 2003). Although it is rural and highly localized, governance of Bragança's mangrove crab fishery (*Ucides cordatus*), and the more recent CT RESEX area, is rooted in a complex history of social-political processes (Partelow et al., 2018a; Saint-Paul and Schneider, 2010).

In Brazil, political efforts to democratize access to natural resources were catalyzed when human-rights and conservation activist Chico Mendes was killed for his efforts in 1988 (Revkin, 2004; Vivacqua, 2018). Mendes, a rubber tapper who believed local communities are the rightful stewards of their environment, argued communities should be given the property rights to manage their land. Following his death, the Reserve (RESEX) program was founded. Today, several Brazilian government agencies including the Chico Mendes Institute for Biodiversity Conservation (ICMBio), the Federal Ministry of the Environment (MMA), and the Institute of the Environment and Renewable Natural Resources (IBAMA) oversee the RESEX policy framework.

The RESEX program attempts to bring often marginalized traditional populations, and those dependent on local natural resources, into national development processes through empowering them to participate in government supported formal co-management processes. Although a core aim is to empower local communities to participate in governance for sustainable resource use, it is also an ambition that local communities, when empowered within a legal framework of common property rights for their shared land, will become conscious stewards of those resources in a way that simultaneously fosters biodiversity conservation (Simonian and Glaser, 2002).

There are few criticisms of the overall RESEX program. The program fits the general shift towards co-management and collaborative governance worldwide (Berkes, 2009; Bodin, 2017). Many view the program as a move in the right direction for environmental justice and an acknowledgement that 'people and nature' governance models (Bennett, 2016; Mace, 2014) are often more effective at achieving both social and ecological goals than traditional 'fortress conservation' models that exclude all human use, or privatization. However, putting ambitious and at times idealistic programs on paper is often followed by long-term difficulties for implementation in practice.

The CT RESEX in Bragança is one of 24 marine RESEX areas in Brazil, most of which have generally struggled to achieve successful outcomes (da Silva, 2004; Santos and Schiavetti, 2014). The SESF is used in this chapter as a tool to diagnose and better understand the complexities of natural resource governance, by examining the variables influencing and interacting to shape outcomes overtime. Because the Bragança area has struggled for more than 20 years to establish effective governance, as reflected on by many locals and organizations involved, an analytical diagnosis can assist in better understanding why.

CT RESEX co-management success requires CA at multiple social levels and within multiple social groups (Table 4.2). Within the legal framework

TABLE 4.2 Stakeholder groups who are members of the RESEX Deliberative Council Each group has one seat on the council (RESEX = marine extractive reserve)

Organization type (GS5)	*Stakeholder group and level of jurisdiction*	*Role/mission*
Public sector	Chico Mendes Institute for Biodiversity Conservation (ICMBio) (Local, Regional, National)	Direct management, implementation and oversight of RESEX areas.
	Camara Municipal of Vereadores (Local)	Legislative body of the municipal administration promulgates organic law and legislative inspection.
	City Council of Bragança (Local)	Seat of executive power, enforces laws.
	National Institute of Colonization and Agrarian Reform (INCRA) (National)	Advanced agrarian reform through formalizing land tenure for economic development.
	Secretary of State for the Environment of Pará (SEMA) (Regional)	Regional office of national environment secretary.
	Institute of the Environment and Renewable Natural Resources (IBAMA)* (National)	Oversees ICMBio, research, permits, resource management and constitutional framework for RESEX.
	Brazilian Institute for Geography and statistics (IBGE) (National)	Collect and analyze geographic information at national level.
	Federal University of Pará (UFPA) (Local, Regional)	University that provides scientific support and consultation.
	Parás Federal Institute for education, science and technology (IFPA) (Local, Regional)	Provides scientific and technological support and consultation.
Community based	Eight community 'Polos' (Local)	Represent communities (each of the eight separately) in the CT RESEX.
	Association of Users CT RESEX (ASSUREMACATA) (Local)	Represents registered users of the RESEX. Holds land title. Implements rules of the management plan.
	Women Movement Paraense Northeast (Regional)	Social feminist movement that has been implementing a network to avoid violations of women's human rights.

Organization type (GS5)	*Stakeholder group and level of jurisdiction*	*Role/mission*
	Union of Artisanal Fishers of Bragança PA (Local)	Represents the rights of artisanal fishers in a local context.
	Breeders and Bragança Beekeepers Association (Local)	Local NGO.
	National Commission for Extractivist Reserves and coastal extractivist towns (CONFREM) (National)	Represent the rights of extractivists in a national context.
	Fishermen cooperative (COOPA) (Local)	Fisher organization.
	Fishing and aquaculture cooperative of the Salgado region (COPESCAR) (Local)	Organize the production and commercialization of fishers.
Hybrid	Enterprise Technical Assistance and Rural Extension Pará (EMATER) (Local, Regional)	Official organization that promotes Technical Assistance and Rural Extension
	Pastoral fishermen commission (CPP) (Catholic Church) (Local, Regional, National)	Promote services for fishers as support for social organization.

*IBAMA is not on the Deliberative Council but oversees RESEX and protected areas throughout Brazil and is tasked with the formal approval and implementation of legislation proposed by the Deliberative Council, through ICMBio.

(constitutional rules), representatives of each stakeholder group are mandated to participate in a Deliberative Council tasked with making decisions about operational and collective choice rules for management within the RESEX boundaries. Much of the Council's role entails decision-making about the fishery, and, to function properly, it requires constructive cooperation among individual representatives on the council, and CA within represented groups participating. The council needs to constructively deliberate and make decisions together. CA and effective representation is also needed within each group, particularly among local fishers and community members, about who will represent them in the council and what their collective interests are.

The focus of this case study is to examine how the variables in the SESF influence CA and cooperation within natural resource governance, and specifically within co-management programs. Although co-management approaches are often championed and recognized as pro-social advancements, they can face considerable difficulties. CA theories would suggest that many RESEX co-management challenges are rooted in a failure to understand the many social and ecological variables that influence how co-management is designed to fit the context in which it is applied, and how and why the ability of stakeholders to work together is influenced by the social and ecological context. As the general goal of RESEX governance is to enable a cooperative arrangement that can better achieve the RESEX goals, the SESF is a useful tool to help identify the underlying variables influencing this. In the case below, we will see that these influences can include the history of existing institutional development and political change processes, the ecological system dynamics as well as the norms and mental models of participating stakeholders, which can strongly influence how and why they cooperate.

Diagnosis of variables influencing co-management

The results below originate from qualitative data collection methods based on key informant interviews and a qualitative content analysis of that data using the SESF as a data coding tool. Below, a brief overview of resulting analysis is provided. The text is marked with the second-tier variable codes of the SESF (e.g., RS1, GS2, RU3). Table 4.3 provides a summary of the key variables and how they influence CA in the CT RESEX.

In Brazil, political stability ensued from 1998 to 2015 with continuous investments into social and economic sectors, eradicating extreme poverty and expanding Brazil's lower middle class (S1). However, many of Brazil's rural natural resource dependent citizens still lack access to higher education and economic opportunities (S2). As of recently, increasing political instability and cuts in educational, environmental, and public service expenditure have had negative implications for the incomes and livelihood chances of the poor (S3). Furthermore, expanding seafood markets in Brazil have turned the Bragança region into a national seafood supply center (S5) with better crab processing technology (S7).

The CT RESEX area has clear biophysical boundaries (RS1) between the Caeté and Taperaçú rivers on the Bragança Peninsula (RS2), approximately 420 km^2 (RS3).[1] However, the mangroves are difficult to navigate, requiring local ecological knowledge to access the deep muddy trails and tidal streams only accessible during high tide (RS4). Carrying large sacks of captured crabs is difficult, so most fishers don't go far into the mangrove (RS9) (Thies-Albrecht, 2016). Seasonality affects cycles of ecosystem functioning (RS7). The rainy season from January to May is more difficult and dangerous for fishing activities than the dry season from June to December. Furthermore, mangrove estuaries are vulnerable to sea-level rise (ECO1). Observations since 1972 have shown mangroves in the area shifting inland, linked to increases in mean sea level (Lara et al., 2010). Mangrove resilience can be compromised by human activities (e.g., fishing, altered water flows, erosion, deforestation) that disturb sedimentation processes (Krauss et al., 2014; Woodroffe et al., 2016).

Mangrove crabs (*Ucides cordatus*) are hardly mobile in their adult lives, typically foraging within a one-meter radius of their burrow (RU1; (Diele and Koch, 2010)). They only go further to reproduce during the 'andada' mating period, six days per month from January to March. Mangrove crabs have a slow growth rate and reproduce during the rainy season peaking January-February (RU2). Females carry eggs for three–four weeks before releasing them during spring tides, spreading them across the estuary (RU1; RU7; (Diele and Koch, 2010)). Fishers distinguish male and female crabs by differences in body shape, reproductive organs, hairiness, and distinct tracks in the burrow (RU6). Mangrove production, beneficial soil bacteria, and crab foraging are interdependent, increases in one create direct or indirect increases in the other two (RU3; RS6; (Koch and Nordhaus, 2010)). Crab density in Northern Brazil is high, ~1,650,000 individual crabs per square kilometer in a healthy forest (RU5; RS5; (Diele and Koch, 2010)). Fishers can catch up to 300 crabs per day, averaging 150 (RU5; (Nascimento et al., 2015)). Crabs can be sold live, cooked or as processed meat. Live crabs are typically sold to patrons (middlemen) for 0.35–0.85 Brazilian Reais per crab (~US$0.1-0.25); processed meat is sold for 8–17.00 Reais (~US$2.25–5.50) per kilo locally depending on the season and quality (RU4), and 60–80 Reais in the state capital Belém.

There are ~5000 families distributed across 50 communities (A1) with the rights to use the CT RESEX (Brazil, 2012). The communities are dispersed around the peninsula, only connected by a central paved road into Bragança (A4). RESEX communities are highly dependent (A8) on extracting local natural resources, are either self-employed or contracted to extract crabs or wood for processing companies. Financial hardships can be directly related to seasonal and other changes in the local availability of crabs and seasonal fishing closures during crab reproduction periods (A8). Socioeconomic conditions are relatively low compared to global development standards and harvesting is often for subsistence, with little access to many public services and infrastructure for some, but not all communities (A2). Fishers tend to harvest locally on foot or close to the coastline with boat access (A3; (Thies-Albrecht, 2016)). Local knowledge on tidal flows, estuary navigation, and crab identification is very important (e.g., tides, burrows, gender, size) (A7). Fishers with boats have greater access to distant mangrove areas, but also incur higher time and monetary costs. Collecting crabs only by hand is legal, but some permits for small net traps are given. However, using poles with large fixed hooks is becoming more common (A9). Hooks can injure undersized or female crabs, leaving higher nonselective mortality rates. Crabs are sold live to patrons (middlemen) or cooked and processed by women in households for sale. More recently, some are processed in local artisanal facilities. Although some processing businesses exist, there is potential for entrepreneurship to establish more fisher independence from patrons (A5) because most fishers are locked into patron-client systems for market access, which are often exploitative (A4; A6; I4).

A 50-year land concession and co management rights to the Caeté-Taperaçú RESEX were given to the registered association of users (ASSUREMACATA) and neighboring communities in 2005 (GS1; GS4). The co-management rights apply exclusively within the RESEX boundaries (GS2), to grant access, resource extraction, management, and exclude others, but the alienation rights and the actual land title are held by the Brazilian state (GS7). However, despite aims to inform the RESEX initiative with assessments since 1998, and the initiation in 2005, official assessments did not begin until 2009 (GS10). The final management plan was only published in 2012. Co-management is centered on the Deliberative Council (DC). The DC is facilitated by ICMBio and comprises local, regional, and national actors (Table 4.2). The DC decides on rules or changes to rules (GS6), and member organizations are considered to be the rule makers (GS5). The management plan was generated and approved by ICMBio and the DC. The operational rules for crab fishing focus on user rights (see Appendix 3 in Partelow et al., 2018a).

A monitoring program was created by IBAMA in which some residents received training (IBAMA, 2005). However, the program was largely unsuccessful because of conflicts with monitoring and rule compliance involving relatives and friends of volunteer agents (Amaral et al., 2008) and was discontinued in 2013 (IBAMA, 2013). Now, although graduated sanctions exist, they are no longer enforced in practice by the users themselves or by external authorities. Nonetheless, registered users are obliged to pay a tax to the DC for monitoring and enforcement. The CT RESEX program was subsidized with development aid (offering new houses and consumables via the Bolsa Verde program), targeting resource dependent households. To qualify, families must be registered residents with ICMBio and earn less than 70 Brazilian Reais (~US$22) per family member per month. The program also provides 300 Brazilian Reais (~US$95) every two months with the aim to reduce overharvesting due to financial needs. Approximately 3,700 families are part of the program, increasing their income significantly. However, criticism for the Bolsa Verde program is that it only focuses on terrestrial resources, which may result in greater pressure on marine resources because of the displacement of terrestrial harvesting (Kasanoski, 2016).

Key interactions influencing CA

Today, a major political barrier is the lack of capable individuals willing to invest time and effort into RESEX politics or to take community leadership roles. This contrasts the period from 1996 to 2005, where political momentum was high (Glaser et al., 2010; Glaser and da Silva Oliveira, 2004). This could be attributed to a lack of lost income compensation for those who volunteer in RESEX politics, contention between leaders over institutional power, a lack of long-term leaders or the minimal participation by young people. A lack of financial incentives or perceived benefits have been stated as key drivers, making fishing instead of political investments a better deal for most. Recent programs to build leadership capacity organized by UNESCO and Rare (an NGO) have attempted to address this challenge, ending the period that associated RESEX membership with subsidized housing.

Communication about the purpose of RESEX has been a continual problem. Many started to believe that RESEX was a government social aid program due to coinciding benefits such as free houses, green scholarships, and numerous consumer goods (e.g., fridges and stoves). As RESEX aligned with numerous social development programs to achieve social goals, many moved into the area only to gain social aid, skewing the public perception of RESEX as exclusively a government hand-out program benefiting registered users. Thus, social aid, in part, undermined the main goal to empower local resource users by making them accountable for management. Instead it created expectations that development aid would be provided regardless of local CA by the users themselves. This caused further dismay when social aid stopped, and many became skeptical of RESEX for not delivering promised benefits. For many RESEX users, there was no need to invest political energy when the perceived benefits were given to them without effort anyway. There was a lack of communication about the original RESEX intentions built on social energy for the area, which are the opportunities for long-term political empowerment over resource management decisions. Difficulties in communication on the original idea of the RESEX was a common theme, as the real purpose shifted in the public perception and, in part, in practice with subsidy programs.

The market prices for mangrove crab have been increasing over time. However, the economic benefits are not being passed on to fishers. Patrons control market access for most remote fishers who cannot transport crabs to market. Patrons have the power to offer low prices, increasing their own gains. There are few patrons to choose from, leaving many fishers with no choice but to sell crabs at low prices. The only way fishers can increase their earnings is to catch more crabs. This can lead to overharvesting, which leads to expanding the areas for fishing, which in turn, can create conflicts with other fishers over fishing grounds or the need to harvest smaller (illegal) crabs locally. Smaller crabs get lower prices, requiring larger quantities to maintain stable income. These interactions create a negative feedback loop, often referred to as a social-ecological trap (Boonstra and De Boer, 2014; Cinner, 2011; Glaser et al., 2010).

Similar to the above interactions, the lack of compensation for lost income during seasonal fishing closures leaves fishers with few alternative income sources. In general, knowledge about the closure rules is low, and a lot of fishing still continues to earn money. When income is low, the mangrove acts as a form of insurance for many families who harvest local resources for subsistence (such as crabs and wood) when they cannot purchase additional food or supplies due to a lack of income from fishing. This can lead to overharvesting of both crabs and non-fishing resources during the seasonal closures, undermining ecosystem resilience, which in turn undermines the ability the same ecosystems to act as a reliable source of food security and insurance that can support social welfare. This is also a negative feedback loop and a form of poverty trap (Barrett, 2008; Barrett et al., 2011).

Although the idea of co-management and political self-empowerment receives little criticism, the transition to making it work in practice is difficult. Transitioning from a situation of no formal governance (e.g., no written rules for fishing or formal processes for decision-making) to co-management, is an institutional novelty for most residents. Although an institutional framework was prepared from 1999 to 2005 to develop the RESEX, many were not used to following rules or participating in the activities and processes of politics needed to make the program work. Politics can be socially messy and often require a strong degree of social organization through leadership, communication and contentious deliberation. Many residents and users are not used to doing these activities, let alone initiating amongst themselves. It is not that they cannot do it or are not willing to do it, the barriers are more fundamental. Many simply do not have a mental model of operating their livelihood within a formal institutional framework. Thus, those who have created and are used to operating within the formal institutional framework, will have more power to use it for their own benefit, simply due to familiarity with it. This can create power asymmetries or further marginalization of local resource users due to a mismatch between different modes of institutional familiarity. For example, it was often stated that although RESEX residents have the largest number of seats on the Deliberative Council, their influence is disproportionately small for this reason (Narahara, 2014; Silva-Junior, 2013). For local RESEX users, institutional learning will likely take time through iteration, communication, trial and error. What is viewed as short-term failures may be part of longer-term successes in adapting and changing formal and informal institutional norms in a way that works for all involved. As it is controlled today, the CT RESEX is now largely steered by politicians and administrators.

Outcomes in the CT RESEX

Concluding the case analysis, we can see that the number of variables influencing the CT RESEX is high (Table 4.3). The SESF has helped identify the individual variables and unpack their interactions and influence on the overall system with collective action theory. Results suggest that policy solutions to natural resource

TABLE 4.3 Second-tier variables with hypotheses within collective action theory (Ostrom 1990; Poteete et al. 2010; Cox 2014a). Analyzed case study values are shown for the Caeté-Taperaçu (CT) RESEX. A general trend is shown for current case values, and a brief contextual explanation. RESEX = marine extractive reserve

SESF	*Theoretical claim/ hypothesis for collective action (CA)*	*Case value*	*Case trend*	*Case explanation*
A5	Accountable leadership increases the likelihood of CA.	Low	Unclear	New resource access avenues precipitated leadership struggles
A6; I2; GS3	Communication increases the likelihood of CA.	Low	Unclear	Geographical isolation, although local leaders communicate via WhatsApp. Procedural rules for participation and timing biased against fishers

SESF	*Theoretical claim/ hypothesis for collective action (CA)*	*Case value*	*Case trend*	*Case explanation*
GS6; I4	Conflict resolution mechanisms increase the likelihood of CA	Low	Unclear	Lack of regular Deliberative Council meetings. Lack of monitoring and rule enforcement.
GS6	External sanctions can override pro-social motivations, decreasing the likelihood of CA.	Low	Unclear	Minimal self-monitoring or external sanctioning occurs.
A2	Cultural heterogeneity decreases the likelihood of CA.	Medium	Stable	Informal social and cultural institutions differ between actor groups on Deliberative Council, but relative homogeneity within ASUREMACATA.
A2	Economic heterogeneity of actors increases the likelihood of CA.	High	Stable	Most RESEX member residents have very low income. Numerous non-RESEX members are wealthier and can invest more
A1	Smaller groups increase the likelihood of CA, reducing transaction costs.	High	Increasing	~5,000 registered families (2009). Many other actors are involved (e.g., NGOs, political, academics, state).
A2; A8	If actors have a common interest, CA is more likely.	Medium	Unclear	Undermined by shifting perceptions and motivations to participate in RESEX.
A8	High dependence on local natural resources can motivate CA.	High	Stable	Residents are highly and directly dependent on local resources, but other RESEX actors less so.
A6	Past collaborations increase the likelihood of social capital and therefore CA.	Medium	Decreasing	Early social-political momentum initiated collective efforts, but this diminished, and past leaders were excluded.

(Continued)

SESF	*Theoretical claim/ hypothesis for collective action (CA)*	*Case value*	*Case trend*	*Case explanation*
A2; A6	If costs bring proportional distribution of benefit, CA is more likely.	Medium	Decreasing	Short – and medium – term benefits are decreasing; long-term benefits are difficult to prioritize and incentivize.
I2; I7, I8	If transaction costs of CA are high, it is more difficult.	High	Unclear	Participation difficult with institutional differences. Long-term nature of benefits reduces short-term incentives. RESEX members are only actors who lose income to attend meetings (e.g., lost fishing time).
A1	If user group boundaries are clear, CA is more likely.	High	Stable	RESEX has clear user group boundaries. Users need to register to receive benefits.
GS6; GS8	Graduated sanctions raise compliance and trust in institutions for CA.	Medium	Unclear	First verbal or written warning, then 30–90 – day suspension of extraction rights, then permanent exclusion. However, no cases of enforcement.
GS6.2	Collective choice rules for decision-making increases the likelihood of CA.	Medium	Unclear	DC is basis for collective choice arrangements. Legal framework exists, but politicians and administrators have largely taken over compared to pre-2005.
GS6.3	Rules fit (e.g., accepted by) outside authorities increases the likelihood of CA.	High	Stable	Constitutionally mandated.

SESF	*Theoretical claim/ hypothesis for collective action (CA)*	*Case value*	*Case trend*	*Case explanation*
GS4; GS8	Governance fit to local needs and institutions increases the likelihood of continued CA.	Medium	Unclear	RESEX is a step in the right direction. Co-management aims to fit and evolve from local informal institutions. Initial social energy has declined, leaving an unclear path forward. Current rule-making mostly excludes local users, current leadership is not local but from the landless movement, in line with shifting priorities of most RESEX members.
GS9	Nested levels of governance increases the likelihood of continued CA.	High	Stable	See Table 4.2
RS2	Clear biophysical boundaries increase the likelihood of CA.	Medium	Stable	Boundaries to land and sea are clear but adjacency to other (mangrove) RESEX are difficult in practice, as the mangrove ecosystem is continuous.
RS3	Moderate biophysical size is more conducive to CA.	Medium	Stable	Communities are scattered with isolated fishing areas, but social interactions are possible.
RS5; RU2	Productivity is curvilinear for CA; too high or low would decrease the likelihood.	High	Decreasing	Crab catches still high, but catch per unit effort increasing, giving a false impression of stability and need for CA among fishers.

(*Continued*)

SESF	*Theoretical claim/ hypothesis for collective action (CA)*	*Case value*	*Case trend*	*Case explanation*
RS7; RU7	System predictability increases the likelihood of CA.	Medium	Unclear	Crab reproduction is highly predictable, closure rules potentially effective if followed. Other factors like sea level rise decrease predictability.
RU6; I1	Resources with distinctive markings can be harvested or managed more selectively, increasing the likelihood of CA.	High	Stable	Easy distinction of crab gender increases ability to harvest selectively, reducing the extraction of reproductive females.
RU4	Low value may not incentivize CA, but high value can lead to overexploitation that is too fast for institutions to respond.	Medium	Increasing	Fishers receive low value from patrons, which does motivate CA, as they need to spend more time fishing. Overall crab prices are going up. However, institutional responses are slow, taking decades to establish RESEX in practice when houses and scholarships displace original motivations for participation.

governance, and the reasons why they succeed or fail, are difficult to simplify generically. They depend heavily on adapting institutions to fit the context, equity in how that is done and the proper incentives to align stakeholder objectives together. Although established with high social energy, RESEX co-management has faced difficulties adapting as the goals and context shift over time. As a result, it has not motivated sufficient CA to pull fishers out of social-ecological traps or to bring diversified economic opportunities and decrease dependence on local natural resources. Social development programs have made progress on reducing local resource dependence, but this is not because of successful co-management. Capacity building is likely needed to regain trust and social energy through increased communication to local residents that emphasizes the opportunities that can come from collective empowerment. Because there are many inherent system characteristics that reduce the likelihood of CA (Table 4.3), developing the right incentives for all actors to participate would likely be a fruitful priority.

Overall, the CT RESEX can be viewed as positive in the context of Brazilian policy progress. However, the need for critical reflection and real progress remains. The idea for RESEX was founded on the social movement led by rubber tappers where formal co-management was the last piece of the puzzle in their strongly self-organized struggle for empowerment. In contrast, formal co-management is the starting point for many RESEX areas today, and making the approach fit to different local contexts is a challenge of fostering the social movements and self-organization in current RESEX areas that mirrors that of those initially inspired by Mendes and the rubber tappers. Thus, the process of building strong political will for self-organization and empowerment within the platform is not an inherent feature of the system (arguably that only a few actors still view as a core goal), but rather the starting point for many RESEX areas, and its main challenge both socially and politically (Batista and Simonian, 2013; Vivacqua, 2018). To analyze why this process is challenging for developing governance that guides cooperative human behavior towards group goals, requires a theoretical foundation about how and why CA and cooperation can work in practice. Diagnosis with the SESF, checking the influence of each variable, is a useful starting point for doing this.

Directions forward: natural resource governance and the SESF

Although collaborative environmental governance is occurring worldwide (Bodin, 2017), in various forms of co-management (Berkes, 2009), establishing formal policies on paper is only the first institutional step towards making them work in practice. The reality of natural resource governance today is that most cases are a mixture of state regulations, varying degrees of privatizations (e.g., property rights, certifications), nongovernmental organizations, and/or self-organized CAs at the community level. These policies are often mixed and designed to work together on paper, but the reality of implementation and how they interact within social-ecological systems can be more nuanced and multifaceted. We can see this in the CT RESEX. The RESEX policy technically exists, and many stakeholders are involved. This looks cooperative and productive on paper, but other factors outlined above such as the ecological conditions, political history, and local stakeholder incentives shape many of the difficulties for achieving success.

We can see that the SESF and collective action theory are useful analytical tools, but they are not the only ones. Many others are outlined in this book, and other literature (Binder et al., 2013; Cox et al., 2016). The multifaceted nature of natural resource systems often requires an analyst to use a combination of theories, frameworks, or methodologies to get a complete picture (Berkes, 2008; Cash et al., 2006). No single theory or framework will entirely explain how a system works or has evolved over time, but each will provide a different perspective or lens to view it.

Ostrom's SESF can help better understand the governance puzzle, showing the usefulness of collective action theories, interdisciplinarity, and systems thinking. The SESF has also advanced the understanding of diagnostic methodologies in the social sciences and addressed the need to solve the case study comparability problem. However, while the case of the CT RESEX has shown how the SESF can be useful, the framework also has limitations, as does any theory or framework. These limitations and directions forward are discussed below.

Although literature on the SESF is widely cited, few have engaged in critiquing the framework despite evident shortcomings and the need for continued research with constructive critique. Existing critique has followed two directions: (1) on its theoretical assumptions and foundations (Saunders, 2014; Singleton, 2017) and (2) on its ability to be applied methodologically (Cox, 2014b; Leslie et al., 2015; Partelow, 2018; Partelow et al., 2018b; Thiel et al., 2015). Other work has examined the framework's inherent ontology and the potential challenges for developing rules structuring systems into nested hierarchies (Frey and Cox, 2015).

Theory and the SESF

The SESF has theoretical inclusion criteria for its variables (Partelow, 2018). Variables have been added to the framework because they have been shown to influence CA in community-based natural resource management. However, critique has been raised on the scale of external validity for theories of CA. In their focused review, Faysse and Mustapha (2017) highlight that disciplinary distinctions can be observed between institutional economics and sociological/anthropological contributions to collective action theory. The scale and external validity are noted as a key difference, where institutional economics approaches tend to make broader utilitarian claims of global external validity, while sociological claims tend to focus on the local level linked to interrelational symbolic or power perspectives within social groups (Faysse and Mustapha, 2017; Mosse, 1997). The SESF arguably, as a result, provides mid-range theories of CA, a balance between locally grounded empirical research and the more universally framed commonalities between them. This distinction highlights parts of the comparability problem, showing the need to integrate different cases and disciplinary understandings to inform the construction and modification of the framework as an interdisciplinary tool of jointly understood variables.

Modifying the framework (i.e., reorganization, addition, and/or subtraction of variables) was intended in its design (Ostrom 2007, 2009). However, there are no published rules for doing this. There is only some literature suggesting guidelines and providing examples. Two aspects can be considered when modifying the framework. (1) What is the purpose of the modifications? Is it to continue building a theoretical framework for analyzing CA in community-based natural resource governance? Or, for example, a different goal may be to identify all variables that exist in the system, characterizing it as a social-ecological system without considering the influence each variable has on CA. (2) When modifying the framework, the system level, context, and/or external validity for which the suggested modifications apply should be explicitly stated. For example, if suggested modifications are derived from research on a small-scale fishery research in Kenya, are the authors suggesting the modified framework then applies to all fisheries in Kenya, or even outside Kenya? Or to forestry cases as well? Both of these aspects relate to methodological transparency when applying the framework, the moving further in trying to solve the comparability problem.

The SESF has many variables, but some aspects of natural resource governance still may not be included. For example, power is not explicitly considered as a variable within the SESF, although it could be argued that it is implicit within variables such as Network structure (GS3), Norms (trust-reciprocity)/social capital (A6) or Knowledge of SES/mental models (A7), among others. There is a general lack of literature within commons and CA scholarship addressing issues of power within social systems and schools of thought arguing the social construction of reality, particularly those that combine disciplinary understandings, beyond a relatively small subset of literature (e.g., Agrawal, 2005; Dahal et al., 2014; Ratner et al., 2013). Power can be difficult to grasp as a concept within governance literature, in part because different schools of thought view it in different ways, as is highlighted by Faysse and Mustapha (2017) in CA research. Qualitative social constructionist approaches within sociology and anthropology may be difficult to integrate with the positivistic generalizations trying to be made by institutional economists. If power influences many aspects of the system, where would the 'power' variable be located ontologically in the system of nested hierarchies in the SESF? Furthermore, framing a social-ecological system as a system of hierarchies may itself be critiqued as a construction of framing the human–nature relationship, one which dictates what we view as worth knowing and what type of knowledge is generated and used to inform governance practice. Nonetheless, power as a general concept is recognized by scholars in many fields as an important piece of understanding institutional development and change processes. Better understanding of how institutions shape interpersonal relationships between individuals and groups, is on-going work. These aspects of natural resource governance are lacking in current versions of the SESF. Scholars from different disciplinary understandings are needed to contribute to the literature on the SESF, enhancing its usefulness as a governance analysis tool.

Similarly, critique has been voiced for the need to include broader ecological criteria and theories through engagement with the SESF from natural scientists (Epstein et al., 2013; Vogt et al., 2015). The SESF was almost exclusively developed by social scientists, and the ecological variables are focused on far less frequently in existing studies (Partelow, 2018). If the SESF aims to be a more well-rounded tool for analyzing governance, further engagement by interdisciplinary scholars that can bring a combined natural science perspective would be useful. The influence of institutional economic thinking is also seen, more discursively, in the framework's terminology. First-tier variable terminology reflects a rather utilitarian view of nature (i.e., Resource Systems and Resource Units). Framing a social-ecological system in this way represents a specific disciplinary epistemology, what is worth knowing about the system and why, a view that may evolve with more interdisciplinary engagement from the natural sciences, and diversity within the social sciences.

Methods and the SESF

The SESF provides an initial list of variables; however, less focus has been put into developing data collection and analysis methods for measuring all of the 56 multidisciplinary second-tier variables in a way that can support a combined analysis. Procedures for doing empirical research with the framework, or 'operationalizing' it, still need to be tested in different contexts to see what methods are useful (Cox, 2014b; Leslie et al., 2015; Partelow et al., 2018b). It can be argued that the next step in solving the comparability problem is increased methodological transparency. Reviews have shown that methodological heterogeneity in studies applying the framework is high (Partelow, 2018; Thiel et al., 2015). This suggests that the framework's variables can be easily adapted to context, a positive aspect, but that having common variables is not enough to allow for comparison across diverse cases. An intermediary step for solving this problem is transparent communication of how variables are adapted to context (e.g., defined) along with data collection and analysis methods used to understand them. Developing a database of variable definitions and methodologies used to measure and analyze them across contexts would be useful. Numerous databases exist, partly addressing this issue (i.e., SESMAD; SESGO; SES Library; CPR Database; NIIS Online Database; IFRI Database). An additional step might be the evolution of sub-frameworks, ones that evolve and adapt the variables and methodologies to specific resource use sectors (e.g., separate contextualized frameworks for fisheries and forestry) (Partelow, 2018). Some research has shown that the framework may evolve into different directions, such as a characterization tool for social-ecological systems generally (Delgado-Serrano and Ramos, 2015), or as a tool for knowledge sharing and deliberation (Partelow et al., 2019).

Concluding thoughts

In conclusion, historians and social anthropologists have shown that CA and cooperation have been humanity's secret weapon for success throughout history. More recently, the work of a more diverse field of environmental social scientists has shown that better understanding how and why people work together is critical for developing natural resource governance that leads to more desirable outcomes for both people and nature. Governance can be viewed as the set of strategies societies use to do this. Whether to exchange goods and services, advance human rights or to create social order, governance is the tool we have to guide human behavior towards our desired outcomes through cooperation. When governance falls short or when institutions (rules and norms) are insufficient in incentivizing desired levels of cooperation, this can lead to violence and war in extreme cases, or more commonly, inefficiencies and frustrations in the delivery of goods and services in daily life. Maintaining ecosystem and social system functioning, welfare and diversity, are direct outcomes of interest to natural resource governance. However, what governance is, as the right combinations of rules and norms to guide human behavior, is very likely to differ between cases. Governance needs to fit the context of the problem it is trying to solve. A main message from Ostrom's research was that there are no panaceas for governance, there are no one-size-fits-all solutions to the world's different environmental problems. On the other hand, we need to be able to generalize common themes that emerge from successful cases by better comparing our findings. From Ostrom's perspective, many social and ecological variables can influence how governance evolves to fit a context, or what influences why a governance approach is having difficulties achieving its goals. The SESF was designed to help identify the variables hindering and enabling natural resource governance by using collective action theory and trying to solve the comparability problem.

Note

1 www.icmbio.gov.br/portal/unidadesdeconservacao/biomas-brasileiros/marinho/unidades-de-conservacao-marinho/2107-resex-marinha-de-caete-taperacu

References

Agrawal, A. (2001) 'Common property institutions and sustainable governance of resources', *World Development*, vol. 29, no. 10, pp. 1649–1672.

Agrawal, A. (2005) *Environmentality: Technologies of Government and the Making of Subjects*. Duke University Press, Durham, NC.

Amaral, L., Almeida, O. and McGrath, G. D. (2008) 'Co-Gestão e o Sistema de Monitoramento Voluntário: A Desistência dos Agentes Ambientais Voluntários no Baixo Amazonas'. *IV Encontro Nacional da Anppas.*

Anderies, J. M. and Janssen, A. M. (2012) 'Elinor Ostrom (1933–2012): Pioneer in the interdisciplinary science of coupled social-ecological systems', *PLoS Biology*, vol. 10, no. 10, p. e1001405.

van Assche, K., Beunen, R. and Duineveld, M. (2014) *Evolutionary Governance Theory*. Springer, Cham.

Barrett, C. B. (2008) 'Poverty traps and resource dynamics in smallholder agrarian systems', in A. Ruijis and R. Dellink (eds) *Economics of Poverty, Environment and Natural-Resource Use*. Springer, New York.

Barrett, C. B., Travis, J. A. and Dasgupta, P. (2011) 'On biodiversity conservation and poverty traps', *Proceedings of the National Academy of Sciences*, vol. 108, no. 34, pp. 13907–13912.

Batista, I. M. S. and Simonian, L. T. L. (2013) 'Implicações políticas, econômicas e socioambientais da RESEX Mãe Grande de Curuçá: Perspectivas de desenvolvimento sustentável no estuário paraense?', *Novos Cadernos NAEA*, vol. 16, no. 1, pp. 203–220.

Bennett, N. J. (2016) 'Using perceptions as evidence to improve conservation and environmental management', *Conservation Biology*, vol. 30, no. 3, pp. 582–592.

Berkes, F. (2008) 'Commons in a multi-level world', *International Journal of the Commons*, vol. 2, no. 1, pp. 1–6.

Berkes, F. (2009) 'Evolution of co-management: Role of knowledge generation, bridging organizations and social learning', *Journal of Environmental Management*, vol. 90, no. 5, pp. 1692–1702.

Binder, C., Hinkel, J., Bots, G. W. P. and Pahl-Wostl, C. (2013) 'Comparison of frameworks for analyzing social-ecological systems', *Ecology and Society*, vol. 18, no. 4, p. 26.

Bodin, Ö. (2017) 'Collaborative environmental governance: Achieving collective action in social-ecological systems', *Science*, vol. 357, no. 6352, p. eaan1114.

Boonstra, W. J. and De Boer, W. F. (2014) 'The historical dynamics of social-ecological traps', *Ambio*, vol. 43, no. 3, pp. 260–274.

Brazil. (2012) *Plano de Manejo da Reserva Extrativista Marinha Caeté-Taperaçu*. Volume I. Diagnóstico, Brazil.

Caradonna, J. L. (2014) *Sustainbility: A History*. Oxford University Press, Oxford.

Cash, D. W., Adger, N. W., Berkes, F., Garden, P., Lebel, L. and Olsson, P. (2006) 'Scale and cross-scale dynamics : Governance and information in a multilevel world', *Ecology and Society*, vol. 11, no. 2: 8. http://www.ecologyandsociety.org/vol11/iss2/art8/.

Cinner, J. E. (2011) 'Social-ecological traps in reef fisheries', *Global Environmental Change*, vol. 21, no. 3, pp. 835–839.

Cox, M. (2014a) 'Understanding large social-ecological systems : Introducing the SES-MAD project', *International Journal of the Commons*, vol. 8, no. 2, pp. 265–276.

Cox, M. (2014b) 'Applying a social-ecological system framework to the study of the Taos Valley Irrigation System', *Human Ecology*, vol. 42, no. 2, pp. 311–324.

Cox, M., Villamayor-tomas, S., Epstein, G., Evans, L., Ban, C. N., Fleischman, F., Nenadovic, M. and Garcia-Lopez, G. (2016) 'Synthesizing theories of natural resource management and governance', *Global Environmental Change*, vol. 39, pp. 45–56.

da Silva, P. P. (2004) 'From common property to co-management: Lessons from Brazil's first maritime extractive reserve', *Marine Policy*, vol. 28, no. 5, pp. 419–428.

Dahal, S., Nepal, K. S. and Schuett, A. M. (2014) 'Examining marginalized communities and local conservation institutions: The case of Nepal's Annapurna conservation area', *Environmental Management*, vol. 53, no. 1, pp. 219–230.

Delgado-Serrano, M. M. and Ramos, A. P. (2015) 'Making Ostrom's framework applicable to characterise social ecological systems at the local level', *International Journal of the Commons*, vol. 9, no. 2, pp. 808–830.

Diele, K. and Koch, V. (2010) 'Comparative population dynamics and life histories of North Brazilian Mangrove Crabs, Genera Uca and Ucides (Ocypodoidea)', in U. Saint-Paul and H. Schneider (eds) *Mangrove Dynamics and Management in North Brazil*. Springer-Verlag, Berlin, Heidelberg.

Epstein, G., Vogt, M. J., Mincey, K. S., Cox, M. and Fischer, B. (2013) 'Missing ecology: Integrating ecological perspectives with the social-ecological system framework', *International Journal of the Commons*, vol. 7, no. 2, pp. 432–453.

Faysse, N. and Mustapha, B. A. (2017) 'Finding common ground between theories of collective action: The potential of analyses at a meso-scale', *International Journal of the Commons*, vol. 11, no. 2, pp. 928–949.

Frey, U. J. and Cox, M. (2015) 'Building a diagnostic ontology of social-ecological systems', *International Journal of the Commons*, vol. 9, no. 2, pp. 595–618.

Glaser, M. (2003) 'Ecosystem, local economy and social sustainability: A case study of Caeté estuary, North Brazil', *Wetlands Ecology and Management*, vol. 11, no. 4, pp. 265–272.

Glaser, M., Krause, G., Oliveira, S. R. and Herazo-Fontalvo, M. (2010) 'Mangroves and people: A social-ecological system', in U. Saint-Paul and H. Schneider (eds) *Mangrove Dynamics and Management in North Brazil*. Springer-Verlag, Berlin, Heidelberg.

Glaser, M. and da Silva Oliveira, R. (2004) 'The prospects for co-management of mangrove ecosystems on the North Brazilian coast – Whose rights, whose duties and whose priorities?', *Natural Resources Forum*, vol. 28, pp. 224–233.

Gordon, H. S. (1954) 'The economic theory of a common-property resource: The fishery', *The Journal of Political Economy*, vol. 62, no. 2, pp. 124–142.

Hardin, G. (1968) 'The tragedy of the commons', *Science*, vol. 162, no. 3859, pp. 1243–1248.

IBAMA. (2005) Instrução Normativa IBAMA.

IBAMA. (2013) Instrução Normativa IBAMA.

Kasanoski, D. S. (2016) *Bolsa verde: benefício social e incentivo à conservação ambiental?* Universidade de Brasília, Brasília.

Klein, P. G. (2000) 'New institutional economics', in B. Bouckaert and G. De Geest (eds) *Encyclopedia of Law and Economics*, Volume 1. Edward Elgar, Northampton, MA.

Koch, V. and Nordhaus, I. (2010) 'Feeding ecology and ecological role of North Brazilian Mangrove Crabs', in U. Saint-Paul and H. Schneider (eds) *Mangrove Dynamics and Management in North Brazil*. Springer-Verlag, Berlin, Heidelberg.

Kooiman, J., Bavinck, M., Chuenpagdee, R., Mahon, R. and R. Pullin. (2008) 'Interactive governance and governability: An introduction', *Journal of Transdisciplinary Environmental Studies*, vol. 7, no. 1, pp. 1–11.

Krauss, K. W., McKee, L. W., Lovelock, E. C., Cahoon, R. D., Saintilan, N., Reef, R. and Chen L. (2014) 'How mangrove forests adjust to rising sea level', *The New Phytologist*, vol. 202, no. 1, pp. 19–34.

Lara, R. J., Cohen, M. and Szlafsztein, C. (2010) 'Drivers of Temporal Changes in Mangrove Vegetation Boundaries and Consequences for Land Use', in U. Saint-Paul and H. Schneider (eds) *Mangrove Dynamics and Management in North Brazil*. Springer-Verlag, Berlin, Heidelberg.

Leslie, H. M., Basurto, X., Nenadovic, M., Sievanen, L., Cavanaugh, C. K., Cota-Nieto, J. J., Erismang, E. B., Finkbeiner, E., Hinojosa-Arango, G., Moreno-Báez, M., Nagavarapu, S., Reddy, W. M. S., Sánchez-Rodríguez, A., Siegel, K., Ulibarria-Valenzuela, J. J., Weaver, H. A. and Aburto-Oropeza, O. (2015) 'Operationalizing the social-ecological systems framework to assess sustainability', *Proceedings of the National Academy of Sciences*, vol. 112, no. 19, pp. 5979–5984.

Mace, G. M. (2014) 'Whose conservation?', *Science*, vol. 345, no. 6204, pp. 1558–1560.

McGinnis, M. D. and Ostrom, E. (2014) 'Social-ecological system framework: Initial changes and continuing challenges', *Ecology and Society*, vol. 19, no. 2, p. 30.

Mosse, D. (1997) 'The symbolic making of a common property resource: History, ecology and locality in a tank-irrigated landscape in South India', *Development and Change*, vol. 28, no. 3, pp. 467–504.

Narahara, K. L. (2014) 'Para evitar questão: a elaboração do Plano de Utilização em uma Reserva Extrativista', in N. Esterci, H. O. Sant'ana-Júnior, and M. J. S. A. Teisserenc (eds) *Territórios socioambientais em construção na Amazônia brasileira*. HIFE (Sociologia & Antropologia), Rio de Janeiro.

Nascimento, J. R., Domingues, D. and Barboza, L. S. D. (2015) 'A cadeia produtiva do caranguejo (Ucides cordatus): os desafios frente às pressões do mercado no território da Resex Marinha Caeté-Taperaçu, Bragança-Pará', *Revista Saude e Ciencia on line*, vol. 4, no. 2, pp. 299–309.

Ostrom, E. (1990) *Governing the Commons: The Evolution of Institutions for Collective Action*. Cambridge University Press, Cambridge.

Ostrom, E. (1998) 'A behavioral approach to the rational choice theory of collective action', *American Political Science Review*, vol. 92, no. 1, pp. 1–22.

Ostrom, E. (2007) 'A diagnostic approach for going beyond panaceas', *Proceedings of the National Academy of Sciences of the United States of America*, vol. 104, no. 39, pp. 15181–15187.

Ostrom, E. (2009) 'A general framework for analyzing sustainability of social-ecological systems', *Science*, vol. 325, no. 5939, pp. 419–422.

Partelow, S. (2018) 'A review of the social-ecological systems framework: Applications, methods, modifications, and challenges', *Ecology and Society*, vol. 23, no. 4, p. 36.

Partelow, S., Fujitani, M., Soundararajan, V. and Schlüter, A. (2019) 'Transforming the social-ecological systems framework into a knowledge exchange and deliberation tool for comanagement', *Ecology and Society*, vol. 24, no. 1, p. 15.

Partelow, S., Glaser, M., Solano Arce, S., Barboza, L. S. R. and Schlüter, A. (2018a) 'Mangroves, fishers, and the struggle for adaptive comanagement: Applying the social-ecological systems framework to a marine extractive reserve (RESEX) in Brazil', *Ecology and Society*, vol. 23, no. 3, p. 19.

Partelow, S., Senff, P., Buhari, N. and Schlüter, A. (2018b) 'Operationalizing the social-ecological systems framework in pond aquaculture', *International Journal of the Commons*, vol. 12, no. 1, pp. 485–518.

Poteete, A., Janssen, A. M. and Ostrom, E. (2010) *Working Together: Collective Action, the Commons and Multiple Methods in Practice*. Princeton Universty Press, Princeton, NJ.

Ratner, B. D., Meinzen-Dick, R., May, C. and Haglund, E. (2013) 'Resource conflict, collective action, and resilience: An analytical framework', *International Journal of the Commons*, vol. 7, no. 1, pp. 183–208.

Revkin, A. (2004) *The Burning Season: The Murder of Chico Mendes and the Fight for the Amazon Rain Forest*. Island Press, Washington, DC.

Saint-Paul, U. and Schneider, H. (2010) *Mangrove Dynamics and Management in North Brazil*. Springer, Heidelberg.

Santos, C. Z. and Schiavetti, A. (2014) 'Assessment of the management in Brazilian marine extractive reserves', *Ocean and Coastal Management*, vol. 93, pp. 26–36.

Saunders, F. P. (2014) 'The promise of common pool resource theory and the reality of commons projects', *International Journal of the Commons*, vol. 8, no. 2, pp. 636–656.

Scott, A. (1955) 'The fishery: The objectives of sole ownership', *Journal of Political Economy*, vol. 63, no. 2, pp. 116–124.

Silva-Junior, S. R. (2013) *Participação e relações de poder no Conselho Deliberativo da Reserva Extrativista Marinha Caeté-Taperaçu*.Universidade Federal do Pará, Bragança-PA.

Simonian, L. T. and Glaser, M. (2002) 'Extractive reserves and the question of sustainability in neotropical ecosystems', *Neotropical Ecosystems: Proceedings of the German-Brazilian workshop*, pp. 767–776.

Singleton, B. E. (2017) 'What's missing from Ostrom? Combining design principles with the theory of sociocultural viability', *Environmental Politics*, vol. 26, no. 6, pp. 994–1014.

Steffen, W., Broadgate, W., Deutsch, L., Gaffney, O. and Ludwig, C. (2015) 'The trajectory of the anthropocene : The great acceleration', *The Anthropocene Review*, vol. 2, no. 1, pp. 81–98.

Stoker, G. (1998) 'Governance as theory: Five propositions', *International Social Science Journal*, vol. 50, no. 155, pp. 17–28.

Thaler, R. H. (2000) 'From homo economicus to homo sapiens', *Journal of Economic Perspectives*, vol. 14, no. 1, pp. 133–141.

Thiel, A., Adamseged, E. M. and Baake, C. (2015) 'Evaluating an instrument for institutional crafting: How Ostrom's social-ecological systems framework is applied', *Environmental Science and Policy*, vol. 53, pp. 152–164.

Thies-Albrecht, O. (2016) *Spatial and Temporal Dispersal of the U. cordatus Crab Fishery in Mangroves Near Bragança, Northern Brazil: Fishers as Optimal Foragers and Resulting Management Applications*. University of Bremen, Germany.

Vivacqua, M. (2018) 'Coastal-marine extractive reserves: Reflections on the pre-implementation stage', *Ambiente & Sociedade*, vol. 21, pp. 1–22.

Vogt, J. M., Epstein, B. G., Mincey, K. S., Fischer, C. B. and Mccord, P. (2015) 'Putting the "E" in SES : Unpacking the ecology in the Ostrom social-ecological system framework', *Ecology and Society*, vol. 20, no. 1, p. 55.

Warming, J. (1911) 'Om Grundrente af Fiskegrunde', *Nationaløkonomisk tidsskrift*, vol. 49, no. 3, pp. 499–505.

Woodroffe, C. D., Rogers, K., McKee, L. K., Lovelock, E. C., Mendelssohn, A. I. and Saintilan, N. (2016) 'Mangrove sedimentation and response to relative sea-level rise', *Annual Review of Marine Science*, vol. 8, no. 1, pp. 243–266.

5

ANALYSING INSTITUTIONS TO EXPLAIN THE PRACTICE AND OUTCOMES OF FISHERIES CO-MANAGEMENT

Fiona Nunan

Introduction

In Chapter 1, it was noted that institutions form part of, and shape, governance systems. As 'rules of the game' (North, 1990: 3), institutions may facilitate and constrain decision-making and outcomes; they may both affect and be affected by governance arrangements and practice. Institutions that engage with governance are wide-ranging. They may be found in government policy and legislation, setting out how a renewable natural resource should be managed and who should be involved in making decisions, and within households and communities, where locally agreed rules inform the use and management of natural resources. Institutions that affect decision-making in relation to access to and benefits from natural resources may come from beyond natural resources and include political systems, kinship and gender relations (Nunan et al., 2015).

There are many approaches to analysing institutions in relation to natural resource governance, with common property theory informing analysis of common property regimes in settings including irrigation and forest management (Ostrom, 1990). Such analysis is particularly associated with the application of Ostrom's 'design principles' (see Chapter 1) and the Institutional Analysis and Development (IAD) framework, where external variables, including rules in use, influence an 'action arena' leading to interaction between actors and subsequently outcomes (McGinnis, 2011). This approach to analysing institutions has been categorised by Cleaver (2012) as 'mainstream', referring to approaches that seek to design institutions for natural resource governance, seeking institutions that have the greatest potential of delivering on sustainability. Such an approach, Cleaver (2012) argues, places too much emphasis on formal institutions to the neglect of 'informal' institutions and suggests that effective and meaningful institutions can be designed and then implemented, delivering sustainable management of natural resources. Cleaver (2012) argues that this is rarely the case; instead, institutions evolve through social interaction and norms over time. Institutions do not exist in isolation but interact; processes of 'bricolage', or piecing together, result in altered and new institutions, affecting the practice and outcomes of natural resource governance and the lives of people affected by governance systems.

This alternative perspective has been referred to as 'critical institutionalism' (Cleaver, 2012), characterised by recognising '(i) complexity of institutions entwined in everyday social life; (ii) their historical formation; and (iii) the interplay between the traditional and the modern, formal and informal arrangements' (Hall et al., 2014: 73). This chapter introduces critical institutionalism and the associated practice of institutional bricolage and examines how these perspectives have been applied in analyses of governing renewable natural resources in low-income countries and what insights have emerged. The chapter goes on to illustrate the application of the approach by identifying and analysing institutions that affect the practice and outcomes of fisheries co-management on Lake Victoria in East Africa.

Fisheries co-management was introduced on the lake in the 1990s, initially through slightly different approaches in the three countries bordering the lake, Kenya, Tanzania and Uganda. A lake-wide management plan, supported by the Implementation of a Fisheries Management Plan project, funded by the European Union, facilitated the development and implementation of a harmonised approach across the lake, enabling a shared understanding of the community-based fisheries management structures, Beach Management Units (BMUs). BMUs were formed to work with government at multiple administrative levels in managing the lake. The analysis focuses on the local, or village, level, identifying how socially-embedded institutions, such as kinship, gender relations and social norms, interact with bureaucratic institutions such as village government, fisheries officers and the formal structures of BMUs. The analysis shows how such interactions shape the practice and outcomes of fisheries co-management and how co-management may be affected by a range of institutions. As a result of interactions between institutions, or their 'piecing together', the practice and outcomes of co-management differ between locations and over time.

A critical institutionalist approach to analysing natural resource governance is shown in this chapter to reveal the messy reality of governing renewable natural resources. One governance system interacts with other systems and with institutions formed for other purposes, reflecting the complexity of social and natural systems and the interactions between them. It highlights the influence of power relations in governance systems and how dominant narratives can perpetuate exclusion of less influential actors through existing and new institutions.

Critical institutionalism and institutional bricolage

Critical institutionalism recognises the diversity of institutions that exist in any situation, which may stem from formal or informal origins, though distinguishing between formal and informal can be difficult and subjective. Cleaver (2002) refers instead to bureaucratic and socially-embedded institutions, defining bureaucratic institutions as 'those formalised arrangements based on explicit organisational structures, contracts and legal rights, often introduced by governments or development agencies' and socially embedded institutions as 'those based on culture, social organisation and daily practice' (2002: 13). As indicated by the characteristics of critical institutionalism, the approach recognises that a diversity of institutions affects how a resource is accessed and managed, not all of which may be directly associated with natural resources. Instead, many of these institutions are embedded in social life.

Given the range and number of institutions that co-exist, it is inevitable that they will interact and influence each other and even modify each other or result in new institutions being formed. This process has been referred to as 'institutional bricolage', which Cleaver et al. (2013: 168) define as 'the processes in which people (consciously or unconsciously) draw on existing social formulae and arrangements (rules, traditions, norms, roles and relationships) to patch together institutions in response to changing situations'. Such piecing together of institutions may particularly happen when new institutions are introduced in a setting and, over time, existing institutions interact with and modify the new institution (de Koning, 2014). Cleaver (2002) attributes her development of institutional bricolage in relation to natural resource management to the work of Mary Douglas on institutions. Douglas (1973 and 1987) contented that 'institutions do the thinking' for people, with institutions and decisions rarely resulting from individual rational choice. Instead, institutions facilitate decisions and action and interact with each other as people navigate through situations.

The 'practice' of institutional bricolage is referred to in some literature, suggesting that different approaches can be taken towards this piecing together of institutions. One example of this is given in three forms of institutional bricolage practice identified by de Koning (2014), of aggregation, alteration and articulation. These were identified from research into how local communities organised community forestry in the Amazon Region of Bolivia and Ecuador, with different responses and different outcomes.

Aggregation is said to refer to 'the recombination of newly introduced institutions and locally embedded institutions' (de Koning, 2014: 360), involving locally embedded institutions, such as traditional beliefs or social norms, being 'pieced together with newly introduced institutions' (de Koning, 2014: 363). Alteration refers to 'the adaptation of institutions. This can happen to both newly introduced institutions and locally embedded institutions. Local actors often tweak or tinker with forest regulations to make them better fit their livelihoods or identity' (de Koning, 2014: 360). Finally, articulation involves 'the claiming of traditional identities and culture and the rejection of newly introduced institutions' (de Koning, 2014: 360).

In de Koning's (2014) cases, an example of aggregation involved a community using forest governance reforms to acquire land titles through a forest management plan despite not already having the titles, as was supposed to be the case.

Instead of the community continuing with timber extraction and forest management, they went back to Brazil nut collection, having achieved what they wanted out of the governance reform. In the case of alteration, farmers, whose main livelihood came from cattle, reshaped new forest regulations through local norms and identity, thus avoiding new regulations on timber extraction. The example of articulation was given in the case of the introduction of new forest regulations in Ecuador to an indigenous community. The community interpreted the regulations as a threat to their identity and independent culture and rejected them. Over time, however, there was some acceptance of the new regulations as some members welcomed government support for development and education. These examples show how governance reforms can be interpreted and used for multiple purposes and how they interact with a range of existing institutions, resulting in diverse outcomes.

This categorisation of practices of institutional bricolage has since been applied elsewhere. For example, Faggin and Behagel (2018) also found multiple bricolage processes in Brazil in response to the introduction of 'sustainable forest management', including rejection through the continuation of illegal practices. Their analysis also highlighted the role of actors beyond local communities in engaging with institutional bricolage, implying that it is not only communities that may interpret and interact with governance reforms in a range of ways and for different purposes.

A further example of actors beyond communities engaging in institutional bricolage is given by Funder and Marani (2015), who apply a critical institutionalist lens to investigate how government Environment Officers in Kenya interacted with communities to implement legislation given the constraints they work within. They show how these officers operate as 'bricoleurs', carrying out their work through 'making their own discrete interpretations of rules and procedures, and forming strategic alliances and informal agreements with communities on the use and management of natural resources' (Funder and Marani, 2015: 89). In this example, District Environment Officers utilise existing institutions and create new institutions to deliver realistic and useful outcomes.

Institutional bricolage then can be a way of actors embedding and giving legitimacy to new institutions. Cleaver and de Koning (2015: 5) refer to bricolage as a way of making institutions 'familiar', being made 'to fit with accepted logics of practice and social relationships' and this process of being made familiar legitimises new, though bricolaged, institutions.

Cleaver and Whaley (2018) built on insights from applications of critical institutionalism to interrogate the theory of adaptive governance, which aims to provide a holistic approach to the management of ecosystems (Folke et al., 2005). It has tended to be thought about in terms of 'recipes' of criteria, including institutional arrangements that facilitate flexibility in responding to new information and change (Nunan, 2016). Cleaver and Whaley (2018) critique this 'recipe' approach, arguing that the ingredients are often those associated with a 'generic spectrum of governance' (2018: 3) and that the model adopts too optimistic a view to developing common understandings, forming trust and enabling collaboration. In response to this critique, they suggest that taking a critical institutionalist approach to adaptive governance and co-management can facilitate a more critical understanding of process, power and meaning (Cleaver and Whaley, 2018). This critical understanding involves explaining 'the processes through which environmental governance arrangements emerge and endure', 'how power is embedded in wider societal relations and reproduced in everyday interactions' and how 'attribution of meaning is crucial for legitimizing and sanctioning relationships by relating them to accepted knowledge and familiar socio-political and environmental orders' (Cleaver and Whaley, 2018: 10).

What these examples of analysis using critical institutionalism and an institutional bricolage lens show is that responses to new institutions are shaped by existing institutions and that these interactions affect the outcomes of governance. Such outcomes are not always positive. Haapala et al. (2016), for example, show how existing institutions led to elite capture of Water User and Sanitation Committees in Nepal, with implications for equity and functionality. However, they also found that this capture led to greater sustainability of the committees, particularly financially. Their findings led them to conclude that 'organic bricolage processes' should be supported, so that institutions emerge that are locally-relevant and supported. This, they suggest, could be done by planning for bricolage from the outset and by developing and strengthening capacity of key institutions (Haapala et al., 2016). They do not reject the development of institutional designs, but recommend that bricolage should be facilitated and encouraged, with support from project staff. This plea for facilitating bricolage does not explicitly set out how power relations could be challenged, and elite capture prevented, but evidence from elsewhere suggests that external actors can assist in reducing the potential for harmful elite capture (Persha and Andersson, 2014) and that over time, non-elites may become so involved in management processes that they have a stronger platform from which to challenge domination of elites (Lund and Saito-Jensen, 2013).

Bersaglio and Cleaver (2018) also highlight the potentially negative outcomes of institutional bricolage for some stakeholders. In their analysis of the formation of community conservancies in Kenya, they show how institutions formed through bricolage are 'entwined with micro-level power relationships in everyday life and historical patterns of access, accumulation, and domination in society at large' (Bersaglio and Cleaver, 2018: 468). Through the influence of existing institutions, new institutions are constrained in their potential to deliver on transformative change. Even 'existing institutions' themselves are likely to have been adapted and modified over time by other institutions.

In terms of data collection methods used in reports on research using critical institutionalism as a guide to research design there is considerable emphasis on collection of qualitative data, from a number of different types of actors and sources, to capture diverse experiences, knowledge and perspectives. A number of methods were often used in search of different insights. For example, de

Koning (2014: 361) referred to using 'in-depth interviews, participant observations, group interviews, group exercises and questionnaires'. Some reports on research methods used refer to using ethnographic approaches. Kairu et al. (2018: 79), for example, reported on the use of 'shadowing', which involved 'following participants in the course of their daily duties and recording their activities' in their research into the implementation of Participatory Forest Management in Kenya. Forest officers at multiple administrative levels were shadowed to capture different experiences and challenges in implementing forest policy. A 'multi-level' approach was adopted in other studies too, reflecting the need to capture the experiences and perspectives of different types of actors, with different responsibilities and experiences. Funder and Marani (2015), for example, interviewed officers within the ministry as well as at the district level to generate a richer understanding of the experience and challenges of environmental management. In addition to interviewing a range of actors, including at multiple administrative levels, several authors refer to undertaking data collection at multiple points in time (see, for example, Bersaglio and Cleaver, 2018; Funder and Marani, 2015) to capture different experiences and follow developments in policy implementation. The types of data collection methods used demonstrates the need to generate an in-depth, rich picture, drawing on multiple perspectives and experiences which enable a range of institutions, and their interaction, to be identified and understood. Many reports on the findings are in the form of stories, detailing how situations emerged and were responded to, including how actors drew on a number of institutions and how new, bricolaged, institutions evolved (see Bersaglio and Cleaver, 2018; Cleaver et al., 2013; Funder and Marani, 2015).

This review of literature on the approach and application of critical institutionalism highlights several important characteristics of the approach and insights on the experience of governing renewable natural resources. The approach brings to the fore the diversity and number of institutions that may influence, and be influenced by, governance arrangements and outcomes. Such institutions may be rooted in different systems – they may be formal or informal, or socially-embedded or bureaucratic – and interact with each other. In these interactions, new institutions may be formed, or institutions shaped in new ways. Different forms of institutional bricolage may be practised, with consequences for governance outcomes. Where new governance arrangements are introduced, such as through community-based natural resource management, they do not enter an institutional vacuum. They interact with and are shaped by existing institutions, with sometimes negative outcomes as the new institutions are captured and controlled by more powerful interests. Existing institutions may also influence and bring about the rejection of new governance arrangements.

Taking a critical institutionalist approach to analysing the governance of renewable natural resources therefore involves identifying the plethora of institutions at play in a governance situation, how they interact and why and with what outcomes. The approach requires analysis of power relations and dynamics in seeking to explain why some institutions dominate and perpetuate certain explanations of natural resource degradation and overexploitation. In so doing, the approach may challenge the prevailing solution and offers explanation for who the winners and losers are in practice.

Fisheries co-management on Lake Victoria

Lake Victoria is the second largest freshwater body in the world, covering 68,000 km^2, and is bordered by Kenya, Tanzania and Uganda. The lake has three main commercial fisheries, Nile perch, Nile tilapia and the small, sardine-like fish, dagaa. Nile perch and Nile tilapia were introduced to the lake by colonial authorities in the 1950s and, by the 1990s, a major international export industry had developed associated with the Nile perch fishery. The development of this commercial fishery attracted increasing numbers of fishers and fishing pressure, with more nets and hooks used per boat. With no limit on the number of licenses issued or catch size, concerns arose in the early 2000s that stocks were depleting. In 2015, it was reported that there were around 200,000 fishers (boat owners and boat crew) on the lake and around 1,500 fish landing sites, varying in size and level of development (LVFO, 2015b).

Concern about the reduction in stocks and catch size of Nile perch was accompanied by concern about the prevalence of illegalities on the lake. Illegalities are manifested in the use of illegal gear to catch undersized fish, illegal fishing methods and sale of undersized fish. In addition, a reduction in capacity of government, resulting from spending restraints associated with the Structural Adjustment Programmes of the 1980s, led to a view that government could not tackle these illegalities alone. The perceived prevalence of illegalities and reduced government capacity contributed to the decision to adopt a co-management approach to fisheries management from the late 1990s. The adoption of co-management focused on the formation of BMUs guided by lake-wide and national guidelines. These guidelines set out the process of forming BMUs, through awareness raising and elections of committees, the functions of BMUs and other actors, including local government fisheries officers, in fisheries management, and how the committee should be composed. A BMU was formed at each landing site with at least 30 boats, with some BMUs consisting of more than one landing site, with each committee required to have 9–15 members. The original BMU guidelines required a composition of 30% boat owners, 30% boat crew, 10% fish traders and 30% other (including boat repairers and net makers), and at least 30% of members to be women. In 2017, however, the guidelines were changed in Uganda as a result of dissatisfaction with the performance of BMUs. The revised guidelines give greater dominance to boat owners, requiring that five of the nine committee members are boat owners, one of which should be the chair (DFRMD, 2017). These new guidelines have not been implemented to date and BMUs remain suspended following their dissolution in 2015 by President Museveni during the election campaign.

BMUs work closely with local fisheries staff, most of which are employed by local rather than central government, as government functions have been decentralised in all three countries, though only relatively recently in Kenya, following the revised Constitution of 2010 and passing of the County Government Act in 2012. Some officers are stationed at landing sites, particularly larger sites at which Nile perch is landed, whilst others are stationed at the district/county level or at administrative levels below. As well as supporting BMUs, fisheries staff are tasked with enforcing legislation and carry out water- and land-based patrols with police and sometimes BMU committee members. The contrasting roles can create tensions in relationships between officers and fisherfolk, though also present opportunities in the form of extracting bribes to allow illegal fishing to continue and for fishers to buy back their confiscated gear (Nunan et al., 2018b).

The analysis reported on here draws on research undertaken in 2015 and on previous studies reported on in Nunan et al. (2015) and Nunan (2010). It therefore largely draws on evidence prior to the dissolution of BMUs in Uganda. The research in 2015 involved interviews with boat owners, boat crew and fish processors/traders at a sample of six landing sites in each country bordering the lake, with this sample including sites on the mainland and on islands, targeting different fisheries and being located in different regions within each country. The main purpose of the research was to investigate the personal networks of fisherfolk and how these networks influence attitudes to and knowledge of co-management. The interviews consisted of a questionnaire to capture information on the respondent's personal networks in terms of who they turn to for guidance on their work and occupation and an interview guide with five sections, on: knowledge of BMU structures, activities and performance, compliance and legitimacy; social groupings; occupation, wellbeing and trust; and, future plans in relation to the perceived condition of the fisheries. Convenience sampling was used to select most respondents at the landing sites, whilst purposive sampling was used for the BMU leaders and government fisheries staff.

Bureaucratic institutions

The main bureaucratic institutions associated with fisheries management are the fisheries management directorates in the three countries bordering the lake and fisheries policy and legislation. The fisheries directorates are located within ministries – the Ministry of Livestock and Fisheries in Tanzania, the Ministry of Agriculture, Livestock, Fisheries and Irrigation in Kenya and the Ministry of Agriculture, Animal Industry and Fisheries (MAAIF) in Uganda. Each directorate leads in developing policy and legislation concerning fisheries management, with much of the implementation carried out by fisheries staff either located at, or employed by, decentralised government. Officers are employed by regional (Tanzania), county (Kenya) and district (Uganda) government and lower levels to enforce regulations, inspect fish landing sites and processing plants and support the development of the sector. Fisheries policy in the three countries aims to deliver on multiple objectives, including increasing the contribution of the sector to the national economies, maintaining employment and conserving stocks (Nunan, 2014). Management measures are not, then, always necessarily working towards the same ends. For example, there are no limits to the number of boat licenses that can be issued, with licensing seen as a source of government revenue rather than a mechanism to control access. Policies and legislation give mixed messages in terms of the emphasis on catching fish for export, food security, income and employment, whilst also conserving stocks for sustainability.

Fisheries rules and regulations focus on restricting the capture of young and breeding fish. Certain illegal gears are indiscriminate in what they catch, such as beach seines which sweep everything into the net, whilst monofilament nets are made of plastic and so can be hard to trace. If lost, they keep on 'fishing' when lost and do not decompose. Gill nets of small mesh size are also illegal, as is trading in small sized fish. These rules and regulations are enforced by government fisheries officers, often working with the police on land- and water-based patrols. A further bureaucratic institution, the judiciary, is therefore also involved in fisheries management through making judgements on cases that make it to court.

In addition to the police and fisheries staff, the army and a Special Enforcement Unit have been involved in enforcing regulations in Uganda. A Special Enforcement Unit was initially formed from the mid-2000s, working separately to the Directorate of Fisheries Resources and having no formal connection to BMUs. Members of the Unit were reported not to have been well-trained in fisheries management and were not adequately resourced. Fishing communities attributed the demands for payments from this Unit, whether fishing or trading legally or illegally, to the need for the Unit to cover their costs and make money. The powerful position of the soldiers in the Unit made resistance difficult for fishing communities.

During the research in 2015, and in response to questions on the experience of fisherfolk of illegalities and of government officers on the performance of BMUs, many respondents spoke about corrupt practices, with all occupational groups accused of participating in corruption. Corruption was reported as involving regular payments being made to fisheries staff and BMU leaders in exchange for the ongoing use of illegal practices and payments being made in exchange for the prevention of the removal of, or for the return of, fishing gear and boats. Bureaucratic institutions therefore draw on socially-embedded power relations and relations of trust built up over time to secure income from bribery, thereby facilitating the continued practice of illegal fishing and undermining of fish stocks.

Socially embedded institutions

Several types of socially embedded institutions are explained and analysed, with implications for governance practice and outcomes identified. These 'types' themselves interact but are considered separately to bring out key institutions and implications for governance.

Kinship

There is a mixed picture of the degree of shared ethnicity, or kinship, at landing sites around the lake. Whilst there are certain ethnic groups that dominate the lake fisheries in each country, there is also evidence of many ethnic groups being present at the landing sites, reflecting the attraction of the fisheries for income-generation, particularly for people unable to make an adequate living from agriculture. In Kenya, the Luo tribe dominate the region around Lake Victoria, with a survey on the lake in 2005 reporting 86% of respondents in Kenya being Luo (Nunan et al., 2007). In Uganda, the Buganda dominate the region, reflecting the location of part of the lake in the Buganda Kingdom, and in Tanzania the largest ethnic group represented is Sukuma (Nunan et al., 2007). The personal networks of fisherfolk were found to be fairly homogenous in terms of ethnicity in research conducted in 2015, though Nunan et al. (2018a) note that this may be related to the dominance of certain groups at landing sites rather than choice. Being of the same kin group can be helpful in terms of getting employment, access to credit and in moving between landing sites in search of better fish stocks and higher prices (Nunan et al., 2018a). However, it has also been reported that kinship can prevent or deter fisherfolk from enforcing regulations (Etiegni et al., 2017; Nunan et al., 2018b). This has implications for the functioning of BMUs; they are mandated to play a role in enforcing regulations and preventing illegal practices. However, BMU committee members may be reluctant to report illegal practices, as doing so could result in fractious relations within kin groups.

Gender relations

The fisheries of Lake Victoria are dominated by men. Around 75% of BMU members are men (LVFO, n.d.) and BMU committees are dominated by men, reflecting the occupational composition and majority of BMU members being men. Women within the fisheries of Lake Victoria are generally involved in local processing of fish (e.g. sun-drying dagaa or smoking tilapia) and/or trading fish in local and regional markets. There are of course women working in fish processing plants, but these are employees rather than having their own businesses within fisheries at the landing sites. There are very few women who own a boat and even fewer who go out to fish. There are then strong gender norms on roles within the fisheries. These are reflected in taboos associated with the sector, which tend to be concerned with gender norms and relations, such as women should not step foot into a boat, they should not be present on certain islands and there should be no sexual relations before going out to fish. Gender and power relations also influence the selling of fish by boat owners and crew, with some expecting sex in exchange for giving access to the fish. This practice of 'fish-for-sex' is associated with fisherfolk moving between landing sites and with a higher prevalence of HIV/AIDS than within other sectors (Nunan, 2010).

These powerful gender norms and relations affect interactions between men and women and within groups of men and women that must affect how BMU committees work. The 2015 research gave a mixed picture in terms of the effective participation of women in BMU committees, with some respondents reporting that women were actively involved and were listened to and others reporting that participation was tokenistic and women tended to be silent on BMU committees. The socially embedded institutions of gender and power relations which enable men to dominate public and private lives at the landing sites therefore interact with the bureaucratic institution of the BMU Committee, shaping the outcomes in a way that generally, though not always, leads to tokenistic participation of women in BMU Committees, with little impact of women on decision-making.

Power relations

Relations of power are manifested in multiple ways within fisheries co-management. Power relations can be observed in the way that co-management was designed and implemented and in the relations between government and BMUs, and within trading and employment relations, with implications for fisheries management. The adoption and implementation of fisheries co-management was led by the national governments, working with donor agencies and project staff. Consultation with fisheries staff at decentralised levels of government, members of fishing communities and other stakeholders, such as owners of fish processing plants and non-governmental organisations (NGOs) working in the fisheries sector, was conducted through a series of national and regional workshops. Given the size of Lake Victoria and the fisheries sector, it is likely that relatively few people were consulted and had a say in practice in how co-management might be shaped and implemented. Central government, and actors with whom they already had good relationships, had more influence on the design of, and planning for, fisheries co-management.

Fisheries co-management in the region was developed, at least in part, to broaden the involvement of different stakeholders in management systems. Previous management approaches at the community level had been dominated by male boat owners. Opening up the system to give boat crew and women an equal say in management at the beach level and beyond through BMU Networks was an important objective of the adoption of co-management (Ogwang et al., 2009). However, power relations between government officers and community members, boat owners and crew, boat owners and fish traders and men and women continue to influence fisheries management.

Examples of such power relations include interference by government fisheries officers in elections of BMU Committee members, which may go either way – either interference to prevent fishers engaged in illegalities from taking on leadership roles or enabling them to, so that payments for fishing illegally continue. Interference has included delaying elections, preventing elected committees from being recognised and requiring that new elections take place.

Within labour relations, boat owners are generally viewed as having more power than boat crew (Nunan, 2006), as they set the terms of payment, provide the boat and gears and can let a crew member go at a moment's notice. However, experienced, reliable and skilful crew are valued and many boat owners will provide credit to crew and help out when difficulties are experienced to encourage crew to stay with them. Whilst boat owners then do generally have a higher status within the fisheries, crew are generally recognised and valued for their work and skills. Relations between owners and crew affect fisheries management and outcomes through decisions made on which gears to use, how many and where to fish. Boat owners tend to make the decisions on which gears to use, which may be illegal, but in some cases, this is done in negotiation with, on the advice of, boat crew.

Relations between fishers and traders also affect fisheries management outcomes. Traders include fish agents who have arrangements with specific fish processing factories to collect Nile perch for processing and export and other traders who buy fish to be sold within domestic and regional markets. Some of the fish for local, national and regional markets may be processed first, for example sun-drying of dagaa. Fish agents are in a stronger financial position than other groups at the fish landing sites and often provide credit to boat owners, particularly when fish catches are low. Such credit may be used to buy or repair gears and boats and then tie the owner to the agent, potentially resulting in lower prices for the fish bought by the agent. Local fish processors and traders, many of whom are women, have less access to resources and less power over fishers. They may though offer credit or gifts, including fish-for-sex as described earlier. These trading relations influence decision-making over fishing methods as they provide finance for purchasing gears and, as women fish processors and traders have been excluded from buying large, legal size, Nile perch by the fish processing plants and the prices they offer, they have turned to small, illegal, size and spoiled Nile perch (Medard, 2012).

Institutional bricolage and governance outcomes

Examples of institutional bricolage can be seen in how BMU Committees, and therefore fisheries management at the beach level, remain dominated by more powerful boat owners, with women largely involved on the periphery of BMUs. Patriarchy and wealth enable men to maintain their dominance in fisheries management, particularly fish agents and boat owners as they own the assets and have greater decision-making power over fishing practices. Patriarchy is also seen in the practice of fish-for-sex, with women having to engage in sexual relationships to gain access to fish for processing and trading. This practice reflects, and affects, the status of women at the landing sites and how they interact and engage with fisheries co-management.

Bricolage can also be observed in the continuation of illegalities, despite the implementation of co-management, with labour and trading relations encouraging the use of illegal gears through demand for fish and provision of credit. Illegalities are also facilitated through ongoing corruption, with bribes demanded and offered to enable illegal gears and practices to continue. Nunan et al. (2018b) demonstrate through analysis of data on the practice and impacts of corruption that the illegalities-corruption link undermines co-management through the lack of genuine enforcement, leading to BMUs lacking motivation as well as capacity to act on illegal practices. This in turn has undermined the legitimacy of co-management, with BMUs in particular seen to have failed in reducing illegal practices.

Fisheries staff engage with institutional bricolage in a number of ways. Their roles as government staff involves both working collaboratively with fisherfolk and enforcing regulations. This can be challenging as staff inevitably get to know fisherfolk well, build friendships and relationships of trust. Such trust though can have negative implications for fisheries management as well as positive. For example, trust can facilitate payment of bribes to allow illegalities to continue as well as facilitate cooperation in enforcement. The social and power relations between government officers and fisherfolk affect the interpretation and practice of co-management, with implications for livelihoods and the fisheries.

These examples of bricolage undermine the aims of fisheries management more generally and co-management specifically. Fisheries management is undermined by the continuation of illegal practices, facilitated by relations of power and trust and linked to corruption. The high levels of illegalities have been consistently blamed for declining Nile perch stocks (LVFO, 2015a, 2016). Policy responses to illegalities continue to emphasise the role of enforcement, including by the army and police, aiming to strengthen capacity for 'monitoring, control and surveillance' (MCS). Co-management is seen as peripheral to this approach, with BMU Committees having some role to play in enforcement, including in keeping registers of fishers and gears, but a more militarised approach dominates.

Co-management is also undermined by male boat owners maintaining their capture of fisheries management at the local level, despite endeavours to empower boat crew and women through their participation in BMU Committees. Whilst some committees have been more successful in terms of the active and effective participation of boat crew and women, many have been affected by the continued dominance of boat owners, the migration of boat crew to other landing sites in search of better catches and higher prices and by women constrained by gender norms and relations.

Bricolage can also support and facilitate co-management through relations of trust and cooperation between government staff and fisherfolk. Such relationships are critical to effective collaboration in managing the fisheries. A further positive example can be seen in how at some landing sites, women have used the opportunity of serving on the BMU Committee to have their voices heard and challenge existing gender norms and relations.

Insights from critical institutionalism for natural resource governance

This review of literature on critical institutionalism and associated institutional bricolage, together with application to the case of Lake Victoria fisheries, highlights three important contributions that this approach can make to the analysis of natural resource governance. These are: (1) that a wide range of institutions are involved in governing natural resources, stemming from different sources, not all developed for the purpose of managing natural resources; (2) that institutions interact and affect each other, at times leading to new or modified institutions; and (3) that new institutions, even those that aim to improve livelihoods, may be influenced in such a way by existing institutions that leads to inequitable outcomes, including reduced access to and benefits from natural resources, for some groups of people.

The strengths of critical institutionalism lie in its attention to institutions embedded in society over time and how institutions that are integral to how society works inevitably interact with institutions that seek to govern natural resources. It reminds those seeking to understand governance systems and its outcomes of the plethora of relevant institutions and how they interact, and with what consequences for livelihoods, wellbeing and the status of natural resources.

The concept of institutional bricolage draws attention to how institutions interact, reflecting the complex institutional web present in daily lives. In addition, as we have seen from the examples in the literature review, institutional bricolage demonstrates how and why new institutions may be interpreted in different ways, with interpretation influenced by existing institutions. The occurrence of different interpretations has implications for how people respond to new institutions and how those new institutions may be shaped over time. It is, therefore, challenging to predict how people will respond to new institutions and what those institutions will look like and how they will perform over time and space.

Identifying and analysing institutions, in terms of their roots, diversity and interactions, implies that an in-depth, lengthy approach to data collection is needed, one that involves immersion in a social setting over time. Qualitative methods of data collection and analysis are therefore appropriate that encourage deep understanding. Whilst interviews, participatory research approaches and focus group interviews can be useful, more ethnographic approaches such as observation and shadowing actors are more likely to generate the necessary depth of evidence and understanding. There may, of course, be a range of interpretations of observations on institutions, in terms of what they are and the extent to which they affect other institutions and governance outcomes. This suggests that triangulation and verification of research results is needed, going beyond feeding back on data with respondents but discussing the data in a more interactive way.

References

Bersaglio, B. and Cleaver, F. (2018) 'Green grab by bricolage – The institutional workings of community conservancies in Kenya', *Conservation and Society*, vol. 16, no. 4, pp. 467–480.

Cleaver, F. (2002) 'Reinventing institutions: Bricolage and the social embeddedness of natural resource management', *European Journal of Development Research*, vol. 14, no. 2, pp. 11–30.

Cleaver, F. (2012) *Development through Bricolage: Rethinking Institutions for Natural Resource Management.* Earthscan, London.

Cleaver, F. and de Koning, J. (2015) 'Furthering critical institutionalism', *International Journal of the Commons*, vol. 9, no. 1, pp. 1–18.

Cleaver, F., Franks, T., Maganga, F. and Hall, K. (2013) 'Institutions, security, and pastoralism: Exploring the limits of hybridity', *African Studies Review*, vol. 56, no. 3, pp. 165–189.

Cleaver, F. and Whaley, L. (2018) 'Understanding process, power, and meaning in adaptive governance: A critical institutional reading', *Ecology and Society*, vol. 23, no. 2: 49. https://doi.org/10.5751/ES-10212-230249.

de Koning, J. (2014) 'Unpredictable outcomes in forestry-governance institutions in practice', *Society and Natural Resources*, vol. 27, no. 4, pp. 358–371.

Department of Fisheries Resource Management and Development (DFRMD). (2017) *Guidelines for Beach Management Units in Uganda.* Ministry of Agriculture, Animal Industry and Fisheries, Republic of Uganda, Entebbe.

Douglas, M. (1973) *Rules and Meanings.* Penguin, Harmondsworth.

Douglas, M. (1987) *How Institutions Think.* Routledge and Kegan Paul, London.

Etiegni, C.A., Irvine, K. and Kooy, M. (2017) 'Playing by whose rules? Community norms and fisheries rules in selected beaches within Lake Victoria (Kenya) co-management', *Environment, Development and Sustainability*, vol. 19, no. 4, pp. 1557–1575.

Faggin, J. M. and Behagel, J. H. (2018) 'Institutional bricolage of sustainable forest management implementation in rural settlements in Caatinga biome, Brazil', *International Journal of the Commons*, vol. 12, no. 2, pp. 275–299.

Folke, C., Hahn, T., Olsson, P. and Norberg, J. (2005) 'Adaptive governance of social-ecological systems', *Annual Review of Environmental Resources*, vol. 30, pp. 441–473.

Funder, M. and Marani, M. (2015) 'Local bureaucrats as bricoleurs. The everyday implementation practices of county environment officers in rural Kenya', *International Journal of the Commons*, vol. 9, no. 1, pp. 87–106.

Haapala, J., Rautanen, S. L., White, P., Keskinen, M. and Varis, O. (2016) 'Facilitating bricolage through more organic institutional designs? The case of water users' associations in rural Nepal', *International Journal of the Commons*, vol. 10, no. 2, pp. 1172–1201.

Hall, K., Cleaver, F., Franks, T. and Manganga, F. (2014) 'Capturing critical institutionalism: A synthesis of key themes and debates', *European Journal of Development Research*, vol. 26, no. 1, pp. 71–86.

Kairu, A., Upton, C., Huxham, M., Kotut, K., Mbeche, R. and Kairo, J. (2018) 'From shiny shoes to muddy reality: Understanding how meso-state actors negotiate the implementation gap in participatory forest management', *Society and Natural Resources*, vol. 31, no. 1, pp. 74–88.

Lund, J. F. and Saito-Jensen, M. (2013) 'Revisiting the issue of elite capture of participatory initiatives', *World Development*, vol. 46, pp. 104–112.

Lake Victoria Fisheries Organisation (LVFO). (n.d.) *People in Lake Victoria Fisheries: Livelihoods, Empowerment and Participation. Information Sheet 1.* Implementation of a Fisheries Management Plan, LVFO, Jinja.

LVFO. (2015a) *LVFO Nile Perch Fisheries Management Plan, 2015–2019 (NPFMP2).* LVFO Secretariat, Jinja.

LVFO. (2015b) *Regional Status Report on Lake Victoria Bi-ennial Frame Surveys between 2000 and 2014. Kenya, Tanzania and Uganda.* LVFO, Jinja.

LVFO. (2016) *Fisheries Management Plan III (FMPIII) for Lake Victoria Fisheries 2016–2020.* LVFO Secretariat, Jinja.

McGinnis, M. D. (2011) 'An introduction to IAD and the language of the ostrom workshop: A simple guide to a complex framework', *Policy Studies Journal*, vol. 39, no. 1, pp. 169–183.

Medard, M. (2012) 'Relations between people, relations about things: Gendered investment and the case of the Lake Victoria fishery, Tanzania', *Signs: Journal of Women in Culture and Society*, vol. 37, no. 3, pp. 555–566.

North, D. C. (1990) *Institutions, Institutional Change and Economic Performance.* Cambridge University Press, Cambridge.

Nunan, F. (2006). 'Empowerment and institutions: Managing fisheries in Uganda', *World Development*, vol. 34, no. 7, pp. 1316–1332.

Nunan, F. (2010) 'Mobility and fisherfolk livelihoods on Lake Victoria: Implications for vulnerability and risk', *Geoforum*, vol. 41, no. 5, pp. 776–785.

Nunan, F. (2014) 'Wealth and welfare? Can fisheries management succeed in achieving multiple objectives? A case study of Lake Victoria, East Africa', *Fish and Fisheries*, vol. 15, no. 1, pp. 134–150.

Nunan, F. (2016) *Topic Guide: Natural Resource Governance.* Evidence on Demand, Hemel Hempstead.

Nunan, F., Abila, R., Lwenya, C., Odongkara, K., Onyango, P. and Yongo, E. (2007) *Regional Synthesis of the Socio-Economic Baseline Survey of the Fishing Communities of Lake Victoria.* Implementation of a Fisheries Management Plan project, Lake Victoria Fisheries Organisation, Jinja.

Nunan, F., Cepić, D., Mbilingi, B., Odongkara, K., Yongo, E., Owili, M., Salehe, M., Mlahagwa, E. and Onyango, P. (2018a) 'Community cohesion: Social and economic ties in the personal networks of fisherfolk', *Society & Natural Resources*, vol. 31, no. 3, pp. 306–319.

Nunan, F., Cepić, D., Yongo, E., Salehe, M., Mbilingi, B., Odongkara, K., Onyango, P., Mlahagwa, E. and Owili, M. (2018b) 'Compliance, corruption and co-management: how corruption fuels illegalities and undermines the legitimacy of fisheries co-management', *International Journal of the Commons*, vol. 12, no. 2, pp. 58–79.

Nunan, F., Hara, M. and Onyango, P. (2015) 'Institutions and co-management in East African inland and Malawi fisheries: A critical perspective', *World Development*, vol. 70, pp. 203–214.

Ogwang, V. O., Ikwaput Nyeko, J. and Mbilinyi, R. (2009) 'Implementing co-management of Lake Victoria's fisheries: Achievements and challenges', *African Journal of Tropical Hydrobiology and Fisheries*, vol. 12, no. 1, pp. 52–58.

Ostrom, E. (1990) *Governing the Commons. The Evolution of Institutions for Collective Action.* Cambridge University Press, Cambridge.

Persha, L. and Andersson, K. (2014) 'Elite capture risk and mitigation in decentralized forest governance regimes', *Global Environmental Change*, vol. 24, pp. 265–276.

6

COMMUNITY GOVERNANCE OF COMMON-POOL RESOURCES

Exploring institutional interfaces

Clare Barnes

Introduction

Common-pool resources (CPRs or commons) such as forests, groundwater, pastures, or fishing grounds, are dynamic spaces where multiple conservation, development and livelihoods goals and practices meet. Such resources are often conceptualised as being rivalrous (their use by one user prohibits another from using them) and having low excludability (it is difficult to exclude others from accessing them). Commons scholars contend that these features of CPRs create particular governance challenges to avoid over-exploitation and under-provisioning of the resources. Since the 1980s, commons scholarship has argued that, under certain conditions, resource users are able to *self-govern* CPRs, that is, to craft institutions which avoid over-exploitation and under-provisioning. Elinor Ostrom and colleagues of the Bloomington Workshop at Indiana University have conducted extensive theoretical and empirical research to analyse how communities of resource users are sometimes able to successfully govern their commons (e.g., Baland and Platteau, 1996; Dietz et al., 2003; Ostrom, 1990; Wade, 1988). At the forefront of this analysis has been the institutions – the rules in use (North, 1990) – purposefully crafted by communities, which create incentives for individual resource users to engage in collective action for governing a particular commons resource. Since the 1990s, analytical attention has broadened significantly *beyond* the purposeful CPR oriented community-level institutions of the early literature. Scholars have looked vertically, to include interactions between community-level CPR institutions and institutions at higher conceptual levels (Cox et al., 2010; McGinnis, 2011; Ostrom, 2005), and also horizontally, to other institutions that shape practices at the community level such as rules and norms around gender, religion, age etc. – and how these affect CPR institutions (Agrawal, 2001; Baland and Platteau, 1996; Martin and

Lemon, 2001; Nagendra and Ghate, 2005; Poteete and Ostrom, 2004; Rydin and Pennington, 2000; Sundar, 2000).

In recent years the ontological foundation for analysing such purposeful CPR institutions based on the bounded rationality of resource users has been challenged by *critical* institutionalism (CI) scholars. Whilst concurring that institutions matter (Cleaver, 2012; Johnson, 2004; Jones, 2015), they argue that institutions cannot be purposefully designed to create incentives for resource users to act collectively (Cleaver, 2002, 2012; Cleaver and de Koning, 2015; Jones, 2015). Rather, institutions emerge through a process of bricolage (Cleaver, 2012; Cleaver and de Koning, 2015; Hall et al., 2014; see Chapter 5 in this volume) – the repurposing, modifying or patching together of dynamic institutions in a way that is seen as acceptable. Critical institutionalism scholars posit that CPR institutions can only be understood as being entwined with wider social, political and historical institutions (Cleaver and de Koning, 2015; Gutu et al., 2014; Leach et al., 1999; Mosse, 2006). This perspective presents both a widening of analytical focus beyond the purposefully designed CPR institutions of a main stream institutionalism (MI) approach and a critique of the nature of purposeful institutions. In order to understand the practices of social actors on the commons, CI scholars propose frameworks which allow for multiple, overlapping and repurposed institutions through which resource user practices of engaging with the commons *make sense* to themselves and other members of the community.

This chapter follows Cleaver's (2012) distinction between Ostrom and colleagues' MI with a focus on purposeful institutions, and the emergent CI perspective, with the aim of contributing to debates engaging across both approaches to the study of the commons (Bennett et al., 2018; Forsyth and Johnson, 2014; Hall et al., 2014; Johnson, 2004). I do this by reflecting on how the conceptualisation of institutions in MI can be enhanced through greater engagement with CI literature. My main argument is twofold. First, MI scholarship is increasingly recognising the plurality of institutions which affect CPR governance, through envisaging *non-deliberate* community institutions (such as gender or caste norms through which social practices are enabled or constrained without a specified purpose) as being separate inputs to *purposeful* CPR institutions. That is, *non-deliberate* community institutions are seen as affecting whether communities are more or less likely to be incentivised or able to design *purposeful* CPR institutions. Second, I draw from CI to argue that the interfaces between *non-deliberate* and *purposeful* CPR institutions are more multifarious than simply seeing non-deliberate institutions as *inputs* to purposeful institutions as conceptualised in much MI literature. MI commons scholars can enrich analyses of interwoven purposeful and non-deliberate community-level institutions affecting practices on the commons through closer engagement with CI literature. This can help with answering research questions around why some CPR institutions and practices are maintained, whilst others have been challenged, adjusted, deferred or ignored, and what the implications of this are for resource sustainability and resource users' livelihoods. In this chapter, I explore these arguments further through a case study of forest commons governance in a community in Andhra Pradesh, India.

Conceptualising institutions in mainstream institutionalism: under which conditions will purposeful CPR institutions emerge?

Garrett Hardin's influential work 'The Tragedy of the Commons' in *Science* (Hardin, 1968), published at a time when debates around limits to growth were prevalent, argues that CPR users face individual incentives to overharvest the resources whilst not investing in its long-term maintenance or protection (i.e. under provisioning of the resource). Based on the assumption that individuals would act out of pure self-interest, Hardin posed that individual resource users would not limit their use of the resource as if they did so, their neighbours would still continue to harvest, the resource would ultimately become degraded or collapse and they would have missed out on any short-term benefits from taking their share. It is in the individual's best interest to continue to harvest the resources. This reasoning led Hardin to theorise that 'Ruin is the destination toward which all men rush, each pursuing his own best interest in a society that believes in the freedom of the commons' (Hardin, 1968: 1244). The way out of such a dilemma according to Hardin, was to nationalise or privatise the resources so that the state or market provide rules of resource use (Ostrom, 1990; Ostrom and Cox, 2010).

This reasoning has been strongly contested since the 1980s for being based on the presumption that commons were seen as open access to all users i.e. devoid of rules for their management (Ostrom, 1990; Ostrom and Cox, 2010; Wade, 1988). Dietz et al. (2002: 5) argue that the individual user behaviour Hardin envisaged could sometimes be controlled by 'communication, trust, the anticipation of future interactions, and the ability to build agreements and rules'. Extensive empirical work by Elinor Ostrom and colleagues has been drawn on to argue that where communities of resource users are able to craft institutions – and so the commons are no longer being used as open access – CPRs *can* be managed sustainably by communities. In her influential book *Governing the Commons* (1990: 1) Ostrom argues, that 'communities of individuals have relied on institutions resembling neither the state nor the market to govern some resource systems with reasonable degrees of success over long periods of time'.

Mainstream CPR scholarship broadly adheres to the definition of institutions as 'the prescriptions that humans use to organise all forms of repetitive and structured interactions' (Ostrom, 2005: 3) which includes formal rules and informal customs, norms and taboos that shape repetitive and structured interaction (Ostrom, 2014). Institutions build communication, trust and norms of reciprocity (Cox et al., 2010), uncertainty surrounding other users' resource use and the resource system is reduced (Gibson et al., 2005; Johnson, 2004) and this stimulates collective action for governing resources.

This mainstream CPR literature is often misrepresented as arguing that community CPR institutions are a panacea for all CPR governance, even though the main vein of commons scholarship in the 1980s was directed at exploring the *conditions under which* communities are more likely to act collectively to create successful CPR institutions (see Baland and Platteau, 1996; Andersson et al., 2014; van Laerhoven and Ostrom, 2007; van Laerhoven and Berge, 2011), and how the institutions themselves shape collective action (Cox et al., 2010). Ostrom and colleagues were not arguing that community CPR institutions are always the appropriate governance arrangement, and neither did they argue that communities will always be able to craft durable institutions through collectively acting to agree on access and withdrawal rules (also see Schlager and Ostrom, 1992), organising the monitoring of compliance to the rules and deciding on the sanctioning of rule infractions (Andersson et al., 2014; Schlager and Ostrom, 1992). In *Governing the Commons* (1990) Ostrom studied varied cases of communities of resource users which had successfully managed their CPR over long time periods, drawing from communities managing common resources such as fisheries, meadows, forests and irrigation located in both the Global North and Global South (e.g. Switzerland, Japan, Spain and the Philippines). From this extensive empirical work, she formulated eight Design Principles, seen as characteristics of robust CPR institutions. These were not intended as a blueprint for designing static institutions, indeed Ostrom stressed the institutional diversity of successful community institutions (Ostrom, 2005, 2014), as communities devise and revise flexible institutions to suit their changing needs, and to adjust to wider institutional changes or resource dynamics. The original Design Principles devised by Ostrom in 1990 (see Chapter 1, this volume) have since been reviewed, adjusted, tweaked and supplemented through application in (mostly) case study research (for reviews see Agrawal, 2001; Agrawal and Chhatre, 2006; Andersson et al., 2014; Cox et al., 2010). They are still often employed in case study research of the commons with an underlying premise of refining principles for institutional design (Agrawal, 2001). In Agrawal's (2001) review of the commons literature of the time, he stated that there are likely to be 30–40 factors affecting the successful management of the commons, categorised into four groups: resource system characteristics, institutional arrangements, external environment and group characteristics – further subdivided through later work into economic and socio-political characteristics (Agrawal, 2007; Agrawal and Chhatre, 2006). With a focus on the institutional arrangements, Andersson et al.'s (2014) review found that institution rule making processes, characteristics of the rules, monitoring and enforcement of rule compliance and imposition of sanctions on rule violators were important for success.

Design principle 8 (nested institutions) was later developed by Ostrom into the Institutional Analysis and Development (IAD) framework, which maps institutional relationships across three levels of action arenas. From lower to higher levels, these are: operational level (where individuals make decisions about their direct interactions with resources); collective-choice (where rules are made by representatives, e.g. resource user committees) and constitutional choice (where rules about how rules at the collective-choice level are made e.g. rules on the composition of resource user committees) (Ostrom, 2005, 2014). The action arenas at each level are influenced by institutions, community attributes and biophysical features of the commons, and the outcomes of decisions made in these arenas affect the incentives in the lower level(s) of the framework (Ostrom, 2005, McGinnis, 2011). The focus remains on institutions devised with the purpose of governing CPRs, with other norms seen as being 'attributes of the community', affecting how actors make decisions or strategies (Ostrom, 2011).

Beyond purposeful institutions in MI: the place of non-deliberate institutions

So far, we have seen that MI scholars view institutions as being composed of formal rules and/or informal norms, constituted at different levels, for the purpose of regenerating or sustaining CPRs. However, this is not to say that other non-CPR focussed, non-deliberate institutions within and beyond communities of resource users (such as age, gender, caste, class, tribe and religion) are completely omitted from analyses. Indeed, leading commons scholars (see Agrawal and Gibson, 1999; Kashwan and Lobo, 2014; Ostrom, 2005) have argued against ignoring divisions and multiple interests in communities (or 'romancing the commons' following Imperial and Yandle (2005: 459)). Heterogeneity along different lines has been acknowledged and linked to differing abilities to act collectively and sustain institutions (Paul and Chakrabarti, 2011; Poteete and Ostrom, 2004; van Laerhoven and Andersson, 2013). This includes technical capacity and empowerment of forest users (Pretty and Ward, 2001), community interests (Pagdee et al., 2006), local knowledge of the biophysical conditions and norms of other appropriators (Ostrom, 1998) and social capital (Westermann et al., 2005).

This research line continues, reflected in recent publications in the International Journal of the Commons in which the purposeful nature of CPR institutions is rarely problematized. However, there is increasing recognition of multiple institutions purposefully crafted for the same CPR (Thapa et al., 2018; Tschopp et al., 2018) and non-deliberate community institutions which are mostly analysed as affecting the ability of communities to purposefully craft and sustain CPR institutions (Aguilera, 2018; Alexander et al., 2018; Bennett et al., 2018; Chai and Zeng, 2018). Some authors are also engaging in a more nuanced discussion of the interrelationships between CPR institutions and non-deliberate institutions – where the latter include caste, gender, democratic and political engagement norms, corruption and cultural norms (Bayazid et al., 2018; Fleischman and Rodriguez Solorzano, 2018; German, 2018; Mudliar and Koontz, 2018; Nunan et al., 2018).

Critical institutionalism: fuzzy boundaries between institutions

This conceptualisation of purposeful CPR institutions is questioned by CI scholars. They argue that, rather than being purposeful, institutions are indefinite, non-determined, fluid, unplanned, non-calculated and non-deliberate, embedded in everyday life and histories, and the boundaries between them are blurred rather than clearly delineated (Cleaver, 2012). Lindegaard (2013: 6) distinguishes between MI's 'rational, functional and designable institutions and more organic or 'messy' institutions, informed by social, cultural and historical contexts'. Institutions are seen as multipurpose, rather than the singular CPR governance purpose of commons institutions in mainstream approaches (Cleaver, 2012; Cleaver and de Koning, 2015). They emerge from *bricoleurs,* actors who consciously and non-consciously combine and transform elements of institutions (Sehring, 2009: 66) with the result being a 'patchwork of the new and second hand, and they include: habitual ways of doing things; well-worn practices adapted to new conditions; organisational arrangements invented or borrowed from elsewhere' (Cleaver and de Koning, 2015: 5). As the new arrangement must feel logical and 'fit' in the particular social context, there is attention within CI for how symbols, categories, discourses and authority relations can seep across institutions – lending authority to, and affecting the meaning of, multiple, overlapping institutions (Cleaver, 2012).

Engagement with CI allows for a broader conceptualisation of the interactions between non-deliberate institutions and purposeful CPR institutions to encompass one institution giving meaning to, legitimising or undermining, creating or diminishing trust in, altering or tweaking, establishing feelings of responsibility for, other institutions. The non-deliberate and purposeful institutions intersect/overlap/alter to create these interrelationships and there could be bleeding of meaning across institutions (Cleaver, 2012).

Forest institutions in Andhra Pradesh, India

I illustrate the rich interfaces between institutions through a case study of a forest-dependent tribal community in Andhra Pradesh (AP), India. In tracing the historical institutional processes of forest tenure transition here, we gain an insight into the stickiness of institutions (Boettke et al., 2008) and the institutional building blocks communities – and those supporting communities (e.g. state or civil society actors) – may consciously or non-consciously draw from, repurpose or alter in their institutional practices.

Forest tenure in India is in a period of flux. Since the passing of the Schedule Tribes and Other Traditional Forest Dwellers (Recognition of Forest Rights) Act, 2006 (hereafter the FRA), forest dwellers are able to claim legal recognition of rights to forests and forest resources. This is a significant departure from the centralised control of forest resources instigated during the colonial period and which continued through to the late 1980s (Guha and Gadgil, 1989; Lee and Wolf, 2018). Whilst the National Forest Policy of 1988 had introduced the Joint Forest Management (JFM) Programme (1990) under which areas of forests were jointly managed between a committee of forest users and Forest Department officials, the strong influence of forest department officials in many cases (Kumar and Kant, 2006; Lee and Wolf, 2018) and the fact that land ownership remained with State Forest Departments, means the FRA cannot be seen as fully decentralised governance. The FRA is viewed by many academics, activists and NGOs as representing a clear departure from the status quo of state control of forests (Barnes and van Laerhoven, 2016; Guha and Gadgil, 1989; Kashwan, 2013; Lee and Wolf, 2018), with estimates that 150 million forest dwellers could have the right to claim community rights under the FRA, bringing approximately 40 million hectares of forests under community management (CFR-LA, 2016). As the nodal agency for the FRA, the Ministry of Tribal Affairs (significantly, *not* the Ministry of Environment, Forests and Climate Change under which the Forest Department sits) is responsible for its implementation. In AP, this responsibility lies with the Integrated Tribal Development Agency (ITDA) as the state government department of the Ministry of Tribal Affairs.

The stipulations in the FRA include (i) provisions on the individual, community rights and community forest resource rights (CFRr) that can be claimed, which include access to resources such as non-timber forest products (NTFPs), or water bodies (ii) the process for claiming these rights, from individual or communities submitting claim forms to community Forest Rights Committees for verification, through a District Level Committee and entry in the land registry, and (iii) the empowerment of right-holders for conservation and protection of the land or resources granted (Saxena et al., 2010; see Lee and Wolf, 2018, for an overview of the claim process). FRA implementation has been extremely slow and haphazard across many states, with the granting of CFRr lagging significantly behind individual rights (Bose, 2011; Lee and Wolf, 2018). A coalition of civil society organisations and academics argued in 2016 (ten years after the FRA legislation was passed) that only 3% of the potential CFRr has been realised (CFR-LA, 2016). Across India, civil society organisations have reported that communities submitting claims for CFRr under the FRA have experienced a high rejection rate and long delays in the claim verification process.

I lean on, and build on, previous empirical research conducted with colleagues in 2015 (Barnes et al., 2017) in which we studied a community of forest users (in this case, a clearly delineated Koya Scheduled Tribe village) in East Godavari district where there is a high dependence on forest resources. The Koyas traditionally practiced podu (shifting cultivation) but now engage in agricultural practices, daily wage employment (such as the Mahatma Gandhi National Rural Employment Guarantee Act which guarantees 100 days of work per household per year) and sale of NTFPs – mostly at times of the year when agriculture and daily wage labour are not available. The village consists of 36 households, with a population of 136 (in 2015) living in thatched huts. Eighty percent of the households have small plots of agricultural land (<2 ha) which is used for growing rice, sorghum (cereal), legumes and sesame (Claus, 2015). The dry deciduous forest size is 700 ha, which is mostly on hills surrounding the village on three sides.

The community set up a Forest Rights Committee at *gram sabha* (village council) level in February 2014, with the support of a local Community Based Organisation (CBO) with long-standing presence in the *mandal* (local administrative area) and a Prime Minister's Rural Development Fellow (PMRDF) working for the Integrated

Tribal Development Agency on a two-year contract (hereafter referred to as ITDA fellow). This committee submitted a CFRr claim and received CFRr in May 2014, which led to a public auction of bamboo harvested from a part of their forest under CFRr, just a month later. A second bamboo auction was held in 2016 by the *gram sabha* largely independently of external support, though witnessed by officials.

Initially seen as a successful case of a CFRr claim under FRA due to the presence of a purposeful forest institution with rules for governing NTFPs, and the large bamboo auction, this village presents an interesting case for analysing interactions between purposeful institutions and non-deliberate institutions. This analysis over a three-year time period after the buzz of success of the first bamboo auction has waned reveals how studying the interfaces between institutions can help explain why some forest resource-related practices are maintained, whilst others have been challenged, adjusted, deferred or ignored.

Methods

Data for this chapter is derived from research conducted for a wider project in January till March 2015, as discussed in Barnes et al. (2017), and supplemented with a follow-up research visit in July 2018. In the main fieldwork period in 2015, a detailed household questionnaire of 25% (9) of the total number of households (36) was conducted (following Angelsen et al., 2011), employing purposeful selection to ensure representative clan numbers. Four focus group discussions in the village were conducted to gather data on NTFP collection and maintenance practices, and forest committee and *gram sabha* meetings. Ethnographic fieldwork included many informal discussions with villagers and committee members, observations of discussions and training programmes led by the CBO and ITDA and semi-structured interviews with local FD and ITDA officials, and both external actors working with the village. Working with a local Koya interpreter with social science training helped significantly with developing trust, and ensuring clarity and appropriateness of our questions. Ten forest plot surveys in the areas designated for bamboo harvest, and other parts of the forest favoured for other NTFP collection, provided evidence of NTFP collection and maintenance practices.

The follow-up research visit in July 2018 involved working with a PhD student translator with research methods training. Ethnographic research was conducted including a group meeting with the six remaining forest committee members, unstructured interviews with the CBO and with 15 villagers, each from different households. The purpose of this visit was to understand changes in institutions and forest governance since 2015.

Analysis of institutional interfaces: case study of institutions under FRA

I continue by first analysing the purposeful forest institution at the operational level supported by the CBO and ITDA fellow, as is commonly the focus of an MI perspective. This includes consideration for higher institutional levels as conceptualised in the IAD framework. The subsequent sections of the analysis draw on the rich considerations of interactions between non-deliberate institutions and purposeful CPR institutions from the CI literature through considering: (i) dynamic social norms (around particular NTFPs, gender, and the safeguarding of social relations); (ii) the legitimacy afforded to the purposeful institution and (iii) the legacies of norms of interaction with state and non-state actors.

Purposeful forest institution

Research in 2015 at the operational level of purposeful forest institutions (following Ostrom's IAD framework) revealed that the majority of community members felt the forest to be degrading over time. This was reported through increasing distance travelled to reach the forest from the village over the last five years (78% of respondents) and to a lesser extent decreasing quality of NTFPs over the same period (56% of respondents) (Barnes et al., 2017). In 2014, a forest committee was set up by the community after many years discussing sustainable harvesting of NTFPs and conflict-resolution with the CBO. Ten members (seven men and three women) were elected on the basis of their ability to communicate. Each of the five clans in the village (within the Koya tribe) has a representative on the committee. The purposeful forest institution facilitated by the CBO included rules on outsiders harvesting from the forest, and on bamboo appropriation (such as how and when to cut the bamboo) and provisioning (e.g. mounding activities to protect new bamboo growth). Norms for monitoring the forest have also developed as young men now do this as a group – but only in the forest area designated for bamboo harvesting. The rules were suggested by a forest committee member at *gram sabha* meetings, which are usually attended by around one third of the village population. It was notable that even though the community collected a wide-range of NTFPs throughout the year (households rely on nine NTFPs on average – for subsistence, selling or both – but could name up to 35 NTFPs) and the vast majority displayed knowledge of sustainable harvesting techniques of many NTFPs, the forest institution only crafted rules for bamboo, aside from a general norm to not cut fruit-bearing trees. It appears that knowledge of sustainable NTFP practices, a reliance on NTFPs and a feeling of ownership over the forest has not translated into a forest institution with formal rules or informal norms for NTFPs, beyond bamboo.

Conceptualising the collective-choice level institutions as being the rules and norms of FRA implementation, allows for consideration of both the formal FRA provisions and the overlapping authorities and norms of interactions between CBO and state actors, which affect the operational level discussed above. This interaction of purposeful institutions across levels is exemplified through the occurrence of two bamboo auctions discussed below.

Analysis of the first bamboo auction in 2014 reveals the overlapping authorities of state and non-state actors in forest governance. This was organised by the ITDA fellow, who chose this village to work in due to its perceived high level of organisation and low level of conflict, which many villagers state is due in part to the CBO presence in the area. The auction was seen as a celebration with food and dancing, and villagers spoke of it with pride. It raised 36,111 rupees per household for those who took part in cutting and transporting bamboo (the vast majority of households) and 1,300,000 rupees for a community fund (a provision in the FRA) which was used to install a solar water tap system -again, organised by the ITDA fellow. FD resistance to the FRA claim process and bamboo harvest was strong and persistent, continuing after rights had been granted through the delay in issuing a transit permit for transporting harvested bamboo from the village. There was much local confusion around the provisions in the FRA on issuing transit permissions (constitutional-level institutions) despite the 2012 amendment to the FRA stating that this fell under the rights of a *gram sabha*. The ITDA fellow's strong advocacy for FRA implementation in the village, their organisation of the bamboo harvest, community bank account and solar water tap system, was supported by the remit of the ITDA for FRA implementation. This was an important influence in the ongoing institutional change process across both collective-choice and operational levels, where control of forest resources is shifting from FD to *gram sabha* control.

A second bamboo auction was organised in 2016 by the villagers, making use of the contacts the ITDA fellow had established to arrange for its advertisement, and to contact potential buyers. The *gram sabha* met to discuss which section of the forest to harvest bamboo from and decided on restrictions to the cutting of young bamboo. The men cut the bamboo and the women carried it to the waiting truck, which took it to a clearing to be arranged in lots according to grade. The ITDA Project Officer and Divisional Forest Officer from the FD were present as observers for the bamboo cutting, and for the auction which followed a few weeks later. The committee members felt that the auction itself had been arranged independently, though discussions and conflicting information on the issuance of transit permits for the sold bamboo continued, revealing the ongoing influence of the collective-choice level institutions on the operational level.

In 2017 the chairperson of the forest committee left to study in the nearest large city and even though their title has nominally been taken over by the previous vice chairperson, the forest committee no longer meets. The current chairperson seems reluctant to hold meetings, or arrange a bamboo auction without the previous chairperson being in the village. Indeed, committee members felt that the next meeting and harvest would only occur once the previous chairperson returned to the village. This leads to pertinent questions about the durability and functioning of the purposeful 'successful' CPR institution at the operational level, which I explore further below through turning to an analysis of institutional interfaces between the purposeful forest institution and non-deliberate community institutions.

Overlapping dynamic social norms and purposeful institutions

The cultural significance of Mahua trees to the Koya remains strong. They avoid cutting Mahua trees, as the seeds and flowers are collected, and the flowers can be used to distil alcohol, which is often consumed by men in the afternoon during collection season and at festivals. The beginning of Mahua collection season is welcomed each year with a festival, which is seen as a culturally significant moment in the year. In informal discussions, most villagers would admit to sometimes cutting some trees to clear areas for agriculture, but were adamant that they would not do this for Mahua trees. Whilst there were generally understood norms (introduced by the CBO) of avoiding cutting trees in order to collect NTFPs, exceptions seemed to be permitted for clearing land for agriculture, apart from the case of Mahua. The social and cultural norms around Mahua reinforce this particular part of the forest institution, though it is interesting that we do not see bleeding of this practice into other tree species which are sometimes cut during NTFP collection.

Whilst Mahua is revered and valued by all Koya, the experiences of the bamboo auctions are highly gendered. The men involved in the forest committee largely discussed with pride how bamboo buyers from cities across the state had visited their village, and how they had bartered with them to receive a high price. Women, including the respected elder lady on the forest committee, would discuss how they had made special food and watched the events unfold. The well-worn gendered experiences of engaging with outsiders (especially officials) are reflected and reinforced through the auctions. Women sell NTFPs at the weekly market in the nearby town, but business with banks and officials is conducted by groups of men. It appears that the auction reinforces such norms as it opens up new networks beyond the village for the men in the forest committee – including with newspapers for placing adverts, potential buyers, officials for observing the auction, bank officials for the community funds generated by the sale – and places these men in positions of controlling the sale and subsequent income. The purposeful forest institution can therefore also be seen as a social institution reinforcing the standing of some within the community, to the exclusion of others.

The interactions between social norms and forest monitoring (introduced by the CBO as part of the purposeful institution) are fluid and multifaceted. As the young men in the village already patrolled the forests and village on market day, whilst the rest of the village were selling NTFPs in the market town, it made sense to them to extend this practice to the forest area and to take on this task on a more regular basis. This was widely seen by other villagers as a visible role that young men should uphold in the community. However, the patrolling only extended to the forest area designated for bamboo harvesting for auction, thus excluding the forest area where women collected other NTFPs. Discussions with the young men revealed that they saw the bamboo designated forest as more important to monitor due to the potential financial gain, revealing overlapping and reinforcing gender institutions around both monitoring practices and perceptions of the value of bamboo. By 2018 this patrolling had diminished and was seen as unnecessary as villagers felt few outsiders entered the bamboo areas of the forest. However, in other forest areas where women collect NTFPs, neighbouring villagers were occasionally seen collecting NTFPs, and women were reluctant to sanction them. They reported feeling awkward confronting family or friends from neighbouring villages, as Koya's strong inter-village kinship ties were seen as important to uphold. This points to friction between, on the one hand, the formal forest monitoring institution at village level and gendered, non-static forest monitoring practices and, on the other hand, strong social institutions which are not confined to village boundaries. Through a greater understanding of these frictions we gain a better sense of why some formal forest institutions are deferred, ignored or only partially followed.

Legitimacy of the purposeful forest institution

Since the previous chair of the forest committee left the village, there have been no committee meetings, discussions of forest governance at *gram sabha* meetings or bamboo harvests. The villagers reported that the bamboo is 'wasting in the hills' and therefore the chance of forest fires is increasing. An MI approach may argue that this postponement of the purposeful forest institution was a question of lack of incentive or ability to hold the harvest. Indeed, there may be a lack of incentive from some villagers as there was a high degree of mistrust across many villagers (outside the committee) over how the previous auction's proceeds had been spent, as no visible community development projects had been funded from the income generated by the second bamboo harvest. However, this was not felt within the committee as several male members of the committee who had been involved in organising the second harvest had the skills and networks to arrange a bamboo harvest in the previous chairperson's absence and freely discussed the high expected value from the next auction. Rather there appeared to be a strong obligation felt by committee members to wait for the chairperson to return before organising another auction, which was tied up with notions of maintaining good social relations with the chairperson and other points of friction in the community for which several committee members felt the chairperson's return was needed (such as changing norms amongst the youth towards NTFPs, political parties entering the village and causing divisions, divisions around alignment to external actors (discussed below) and Christian conversions increasing in the area). The purposeful forest institution can again be seen as serving multiple purposes, with its functioning inseparable from wider social institutions. This was supported by the bleeding of authority from the formalised forest institution under the FRA (constitutional level) with its requirement that the signature of the absent previous chairperson (now vice chairperson), as well as the current chairperson, being required to access the community account for funding community development projects.

Therefore, the stalled forest institution and bamboo harvest practices can be usefully analysed as being part of wider frictions in the village, and influenced by norms around dealing with these being centred on the absent chairperson, reinforced by formal FRA requirements for accessing the bank account. Questions of the legitimacy of the forest institution are entwined with the legitimacy afforded to overlapping non-deliberate institutions as well as the formal FRA rules devised at higher institutional levels.

Sticky legacies of the norms around engaging with external actors

The focus on bamboo in the purposeful forest institution can be understood as a reflection of the way the FRA was interpreted by the CBO and ITDA fellow and the meaning they afforded bamboo, which was legitimised through their personal relationships with some members of the forest committee. Whilst the CBO tried to instil a purposeful forest institution for multiple NTFPs, and engaged in training on sustainable NTFP harvesting and processing, their work under the FRA focussed initially on bamboo. The CBO expected that a focus on bamboo would lead to fast, tangible, financial livelihood benefits for the whole community, which in turn would provide incentives for institutions to develop around other NTFPs. The ITDA fellow placed the emphasis on rights-claiming under FRA, seeing the bamboo auction as a community right. The norms of the relationships between the community and these external actors, in which villagers are strongly influenced by the views and actions of the external actors, have significantly affected how forest governance was conceived by the community. In 2015, when reflecting on forest governance in the village, villagers discussed the trainings in sustainable harvesting of multiple NTFPs provided by the CBO, committee meetings and forest monitoring. However, by 2018 the majority of respondents associated forest governance solely with the bamboo auction which seems to have become the acceptable way to discuss the forest. This repurposing of the FRA by both external actors, and subsequent influence in the village, is reflected in this remark by the CBO '*We also influence them* [the villagers] *in how we talk about FRA – they say it is about bamboo because that is what we talk about*'.

Many respondents also reflected upon their perceived need for an outsider to support them, with one older lady reflecting in 2018 '*That lady* [ITDA fellow] *used to roam with the chair – she was the last person to motivate us*'. The ITDA's approach of undertaking the majority of the first bamboo harvest organising herself, and to train a few younger men on the committee, led to tensions with the CBO's longer-term approach (see Barnes et al., 2017) and with some older community members aligned with the CBO. It also meant that only a few community members had been significantly involved in the bamboo auction, and they were even taken to showcase their success at a national level meeting of FRA activists and NGOs. This created an association held by many in the village between the bamboo auction and these few individuals, and also an expectation expressed by many in 2015 that long-term support would be provided by the

ITDA, which had subsequently waned by 2018 due to lack of this expected support materialising. Whilst the entrenched power of the FD in the area (manifest through resistance to FRA implementation) would help to explain the need for such intensive initial support by the ITDA fellow, it does seem to have left legacies in the community. There are still two separate groups in the village formed around the external actors even though the ITDA left the area in 2015 and the CBO presence has significantly reduced.

The long-standing norms of interaction between the Forest Department and villagers create an institutional stickiness with a legacy for newer institutions emerging under the FRA. Under the JFM policy, the community had formed a committee with the FD, where community members faced strict restrictions for taking NTFPs, such as firewood or bamboo for house construction leading to frequent conflicts with the FD (Claus, 2015). The FD manages parts of the forest and plantations beyond the community's CFR area and still employs villagers as daily wage labourers in these areas. There is therefore a legacy of conflictual and contractual institutions between villagers and the FD which affects the meaning afforded by villagers to the forest institution under FRA. The conflictual relationship with the FD is felt by community members who are confused about the permissions they feel they require from the FD in order to hold an auction and transport the bamboo. We could see this as an overbearing non-deliberate institution affecting the trust placed in, and the legitimacy of, the purposeful forest institution.

Conclusion

This chapter aimed to contribute to discussions on the interfaces between purposeful institutions and non-deliberate institutions as conceptualised in Mainstream Institutionalism and Critical Institutionalism literature. I hope to further encourage the recent discussions within MI around decentring purposeful CPR institutions from analyses of commons users' practices. Through the case study analysis, we have seen that there are grounds to question the primacy of purposeful institutions for answering questions around actors' practices on the commons. Indeed, if we want to understand why some institutions are durable, whilst others lose traction, are experienced differently, or are deferred in some way, we can usefully turn to analyses of non-deliberate institutions which appear to hold explanatory power. Whilst an analysis of the purposeful forest institutions across levels offered initial insights into institutions around NTFPs and explanations for the distinction in practices between bamboo and other NTFPs, it is only when we place these institutions within overlapping non-deliberate institutions, that we can understand how different villagers afforded meaning and legitimacy to the forest institution, and how the forest institution is both part of, and reinforces, social norms in the community.

We have also seen that the relationship between non-deliberate and purposeful institutions is far from a simple linear arrangement in which the non-deliberate institutions provide incentives or abilities to devise purposeful institutions.

Rather, we can see complex, entwined institutions existing within legacies of 'past' institutions (e.g. norms around interactions between the FD and communities). For example, in the case study we saw congruence between social norms around male youth protecting the village and the monitoring of the forest, which initially strengthened the perceived appropriateness of monitoring the economically valuable part of the forest. However, we also saw how forest monitoring could be undermined by strong inter-village kinship ties making it inappropriate to sanction neighbouring villagers who break forest rules. Similarly, the way the bamboo auctions were arranged (which were themselves influenced by norms around villagers' engagement with state and non-state actors and the formal provisions in FRA) both reflected and reinforced gender institutions in the village.

Recent MI literature (e.g. Bayazid et al., 2018; Fleischman and Rodriguez Solorzano, 2018; German, 2018; Mudliar and Koontz, 2018; Nunan et al., 2018) shows that a shift away from studying [purposeful CPR] institutions with the goal of 'getting them right', towards a closer understanding of wider social institutions, heterogeneous communities, power relations and practices, is well underway. However, conceptual questions around the compatibility of the ontologically different approaches to institutions remain (Bennett et al., 2018; Forsyth and Johnson, 2014; Hall et al., 2014). This chapter hopes to contribute to these discussions, through engaging an analytical approach which places purposeful institutions within a patchwork of overlapping non-deliberate institutions, showing the value of doing so for answering questions around changing social practices on the commons.

References

Agrawal, A. (2001) 'Common property institutions and sustainable governance of resources', *World Development*, vol. 29, no. 10, pp. 1649–1672.

Agrawal, A. (2007) 'Forests, governance, and sustainability: Common property theory and its contributions', *International Journal of the Commons*, vol. 1, no. 1, pp. 111–136.

Agrawal, A. and Chhatre, A. (2006) 'Explaining success on the commons: Community forest governance in the Indian Himalaya', *World Development,* vol. 34, no. 1, pp. 149–166.

Agrawal, A. and Gibson, C. C. (1999) 'Enchantment and disenchantment: The role of community in natural resource conservation', *World Development*, vol. 27, no. 4, pp. 629–649.

Aguilera, S. E. (2018) 'Measuring squid fishery governance efficacy: A social-ecological system analysis', *International Journal of the Commons*, vol. 12, no. 2, pp. 21–57.

Alexander, S. M., Bodin, Ö. and Barnes, M. L. (2018) 'Untangling the drivers of community cohesion in small-scale fisheries', *International Journal of the Commons*, vol. 12, no. 1, pp. 519–547.

Andersson, K., Benavides, J. P. and León, R. (2014) 'Institutional diversity and local forest governance', *Environmental Science and Policy*, vol. 36, pp. 61–72.

Angelsen, A., Larsen, H. O., Lund, J. F., Smith-Hall, C. and Wunder, S. (2011) *Measuring Livelihoods and Environmental Dependence: Methods for Research and Fieldwork*. CIFOR, Bogor, Indonesia.

Baland, J. and Platteau, J. (1996) *Halting Degradation of Natural Resources. Is There a Role for Rural Communities?* Clarendon Press, Oxford.

Barnes, C., Claus, R., Driessen, P., George, M. A., Ferreira Dos Santos, M. J. and van Laerhoven, F. (2017) 'Uniting forest and livelihood outcomes? Analyzing external actor interventions in sustainable livelihoods in a community forest management context', *International Journal of the Commons*, vol. 11, no. 1, pp. 532–571.

Barnes, C. and van Laerhoven, F. (2016) 'Advocating for change? How a civil society-led coalition influences the implementation of the forest rights act in India', *World Development,* vol. 84, pp. 162–175.

Bayazid, Y., Miyanishi, T., Umetsu, C. and Hamasaki, H. (2018) 'The evolution of a floodplain aquaculture management system in Bangladesh', *International Journal of the Commons*, vol. 12, no. 1, pp. 249–277.

Bebbington, A. (2005) 'Donor–NGO relations and representations of livelihood in non-governmental aid chains', *World Development*, vol. 33, no. 6, pp. 937–950.

Bennett, A., Acton, L., Epstein, G., Gruby, R. L. and Nenadovic, M. (2018) 'Embracing conceptual diversity to integrate power and institutional analysis: Introducing a relational typology', *International Journal of the Commons*, vol. 12, no. 2, pp. 330–357.

Boettke, P., Coyne, C. and Leeson, P. (2008) 'Institutional stickiness and the new development economics', *American Journal of Economics and Sociology*, vol. 67, no. 2, pp. 331–358.

Bose, P. (2011) 'Forest tenure reform: Exclusion of tribal women's rights in semi-arid Rajasthan, India', *International Forestry Review*, vol. 13, no. 2, pp. 220–232.

CFR-LA. (2016) *Promise and Performance: Ten Years of the Forest Rights Act in India. Citizens' Report on Promise and Performance of The Scheduled Tribes and Other Traditional Forest Dwellers (Recognition of Forest Rights) Act, 2006, after 10 years of Its Enactment.* CFR-LA, India.

Chai, Y. and Zeng, Y. (2018) 'Social capital, institutional change, and adaptive governance of the 50-year-old Wang hilltop pond irrigation system in Guangdong, China', *International Journal of the Commons*, vol. 12, no. 2, pp. 191–216.

Claus, R. (2015) 'Making it count: A livelihoods assessment of external actor interventions in Indian Community-Based Forest Management', MSc thesis, Utrecht University, Utrecht, The Netherlands.

Cleaver, F. (2002) 'Reinventing institutions: Bricolage and the social embeddedness of natural resource management', *European Journal of Development Research*, vol. 14, no. 2, pp. 11–30.

Cleaver. F. (2012) *Development through Bricolage: Rethinking Institutions for Natural Resources Management.* Routledge, New York.

Cleaver, F. and de Koning, J. (2015) 'Furthering critical institutionalism', *International Journal of the Commons*, vol. 9, no. 1, pp. 1–18.

Cox, M., Arnold, G. and Villamayor Tomás, S. (2010) 'A review of design principles for community-based natural resource management', *Ecology and Society*, vol. 15, no. 4, p. 38.

Dietz, T., Dolšak, N., Ostrom, E. and Stern, P. C. (2002) 'The drama of the commons', in National Research Council (eds) *The Drama of the Commons.* Committee on the Human Dimensions of Global Change. E. Ostrom, T. Dietz, N. Dolšak, P. C. Stern, S. Stovich, and E. U. Weber, Eds. Division of Behavioral and Social Sciences and Education. National Academy Press, Washington DC.

Dietz, T., Ostrom, E. and Stern, P. (2003) 'The struggle to govern the commons', *Science*, vol. 302, no. 5652, pp. 1907–1912.

Fleischman, F. and Rodriguez Solorzano, C. (2018) 'Institutional supply, public demand, and citizen capabilities to participate in environmental programs in Mexico and India', *International Journal of the Commons,* vol. 12, no. 2, pp. 162–190.

Forsyth, T. and Johnson, C. (2014) 'Elinor Ostrom's legacy: Governing the commons and the rational choice controversy', *Development and Change*, vol. 45, no. 5, pp. 1093–1110.

German, L. (2018) 'Catalyzing self-governance: Addressing multi-faceted collective action dilemmas in densely settled agrarian landscapes', *International Journal of the Commons*, vol. 12, no. 2, pp. 217–250.

Guha, R. and Gadgil, M. (1989) 'State forests and social conflicts in British India: Past and Present', *Journal of Historical Studies*, vol. 123, pp. 143–177.

Gibson, C. C., Williams, J. T. and Ostrom, E. (2005) 'Local enforcement and better forests', *World Development*, vol. 33, no. 2, pp. 273–284.

Gutu, T., Wong, S. and Kinati, W. (2014) 'Institutional bricolage as a new perspective to analyse institutions of communal irrigation: Implications towards meeting the water needs of the poor communities in rural Ethiopia', *Pinnacle Agricultural Research & Management*, vol. 2, no. 1, pp. 1–14.

Hall, K., Cleaver, F., Franks, T. and Maganga, F. (2014) 'Capturing critical institutionalism: A synthesis of key themes and debates', *European Journal of Development Research*, vol. 26, pp. 71–86.

Hardin, G. (1968) 'The Tragedy of the Commons' *Science*, vol. 162, no. 3859, pp. 1243–1248.

Imperial, M. T. and Yandle, T. (2005) 'Taking institutions seriously: Using the IAD framework to analyze fisheries policy', *Society and Natural Resources*, vol. 18, no. 6, pp. 493–509.

Johnson, C. (2004) 'Uncommon ground: The 'poverty of history' in common property discourse', *Development and Change*, vol. 35, no. 3, pp. 407–433.

Jones, S. (2015) 'Bridging political economy analysis and critical institutionalism: An approach to help analyse institutional change for rural water services', *International Journal of the Commons*, vol. 9, no. 1, pp. 65–86.

Kashwan, P. (2013) 'The politics of rights-based approaches in conservation', *Land Use Policy* vol. 31, pp. 613–626.

Kashwan, P. and Lobo, V. (2014) 'Of rights and regeneration: The politics of governing forest and non-forest commons', in S. Lele and A. Menon (eds) *Democratizing Forest Governance in India*. Oxford University Press, Delhi.

Kumar, S. and Kant, S. (2006) 'Organizational resistance to participatory approaches in public agencies: An analysis of forest department's resistance to community-based forest management', *International Public Management Journal*, vol. 9, no. 2, pp. 141–173.

Leach, M., Mearns, R. and Scoones, I. (1999) 'Environmental entitlements: Dynamics and institutions in community-based natural resource management', *World Development*, vol. 27, no. 2, pp. 225–247.

Lee, J. I. and Wolf, S. A. (2018) 'Critical assessment of implementation of the Forest Rights Act of India', *Land Use Policy*, vol. 79, pp. 834–844.

Lindegaard, L. S. (2011) 'Power and change: Locating institutional change theories in a power context', *DIIS Working Paper 2013:20*. Danish Institute for International Studies, Denmark.

Martin, A. and Lemon, M. (2001) 'Challenges for participatory institutions: The case of village forest committee in Karnataka, South India', *Society and Natural Resources*, vol. 14, no. 7, pp. 585–597.

McGinnis, M. D. (2011) 'An introduction to IAD and the language of the Ostrom workshop: A simple guide to a complex framework', *Policy Studies Journal*, vol. 39, no. 1, pp. 169–183.

Mosse, D. (2006) 'Collective action, common property, and social capital in South India: An anthropological commentary', *Economic Development and Cultural Change*, vol. 54, no. 3, pp. 695–724.

Mudliar, P. and Koontz, T. (2018) 'The muting and unmuting of caste across inter-linked action arenas: Inequality and collective action in a community-based watershed group', *International Journal of the Commons*, vol. 12, no. 1, pp. 225–248.

Nagendra, H. and Ghate, R. (2005) 'Role of monitoring in institutional performance: Forest management in Maharashtra, India', *Conservation and Society*, vol. 3, no. 2, pp. 509–532.

North, D. (1990) *Institutions, Institutional Change and Economic Performance*. Cambridge, University Press, Cambridge.

Nunan, F., Cepić, D., Yongo, E., Salehe, M., Mbilingi, B., Odongkara, K., Onyango, P., Mlahagwa, E. and Owili, M. (2018) 'Compliance, corruption and co-management: How corruption fuels illegalities and undermines the legitimacy of fisheries co-management', *International Journal of the Commons*, vol. 12, no. 2, 58–79.

Ostrom, E. (1990) *Governing the commons: The evolution of institutions for collective action*. Cambridge University Press, Cambridge.

Ostrom, E. (1998) 'A behavioral approach to the rational choice theory of collective action', *American Political Science Review*, vol. 92, no. 1, pp. 1–22.

Ostrom, E. (2005) *Understanding Institutional Diversity*. Princeton University Press, Princeton, NJ.

Ostrom, E. (2014) 'Do institutions for collective action evolve?', *Journal of Bioeconomics*, vol. 16, pp. 3–30.

Ostrom, E. and Cox, M. (2010) 'Moving beyond panaceas: A multi-tiered diagnostic approach for social-ecological analysis', *Environmental Conservation*, vol. 37, no. 4, pp. 451–463.

Pagdee, A., Kin, Y. and Daugherty, P. J. (2006) 'What makes community forest management successful: A meta-study from community forests throughout the world', *Society and Natural Resources*, vol. 19, no. 1, pp. 33–52.

Paul, S. and Chakrabarti, S. (2011) 'Socio-economic issues in forest management in India', *Forest Policy & Economics*, vol. 13, no. 1, pp. 55–60.

Poteete, A. and Ostrom, E. (2004) 'Heterogeneity, group size and collective action: The role of institutions in forest management', *Development and Change*, vol. 35, no. 3, pp. 435–461.

Pretty, J. and Ward, H. (2001) 'Social capital and the environment', *World Development*, vol. 29, no. 2, pp. 209–227.

Rydin, Y. and Pennington, N. (2000) 'Public participation and local environmental planning: The collective action problem and the potential of social capital', *Local Environment*, vol. 5, no. 2, pp. 153–169.

Saxena, N. C. et al. (2010) *Report National Committee on Forest Rights Act*. A Joint Committee of Ministry of Environment and Forests and Ministry of Tribal Affairs, Government of India, India.

Schlager, E. and Ostrom, E. (1992) 'Property rights regimes and natural resources: A conceptual analysis', *Land Economics*, vol. 68, no. 3, pp. 249–62.

Sehring, J. (2009) 'Path Dependencies and Institutional Bricolage in Post-Soviet Water Governance', *Water Alternatives*, vol. 2, no. 1, pp. 61–81.

Sundar, N. (2000) 'Unpacking the "joint" in joint forest management', *Development and Change*, vol. 31, no. 1, pp. 255–279.

Thapa, K., Gnyawali, T. P., Chaudhary, L., Chaudhary, B. D., Chaudhary, M., Thapa, G. J., Khanal, C., Karki Thapa, M., Dhakal, T., Rai, D. P., Bhatta, S. R., Upadhya, D., Bhandari, A. R. and Joshi, D. (2018) 'Linkages among forest, water, and wildlife:

A case study from Kalapani community forest in Lamahi bottleneck area in Terai Arc Landscape', *International Journal of the Commons*, vol. 12, no. 2, pp. 1–20.
Tschopp, M., Bieri, S. and Rist, S. (2018) 'Quinoa and production rules: How are cooperatives contributing to governance of natural resources?', *International Journal of the Commons*, vol. 12, no. 1, pp. 402–427.
Van Laerhoven, F. and Andersson, K. (2013) 'The virtue of conflict: An institutional approach to the study of conflict in community forest governance', *International Forestry Review*, vol. 15, no. 1, pp. 122–135.
Van Laerhoven, F. and Berge, E. (2011) 'The 20th anniversary of Elinor Ostrom's governing the commons', *International Journal of the Commons*, vol. 5, no. 1, pp. 1–8.
Van Laerhoven, F. and Ostrom, E. (2007) 'Traditions and trends in the study of the commons', *International Journal of the Commons*, vol. 1, no. 1, pp. 3–28.
Wade, R. (1988) *Village republics: Economic conditions for collective action in South India*. ICS Press, Oakland, CA.
Westermann, O., Ashby, J. and Pretty, J. (2005) 'Gender and social capital: The importance of gender differences for the maturity and effectiveness of natural resource management groups', *World Development*, vol. 33, no. 11, pp. 1783–1799.

7

INDIGENOUS POLYCENTRIC AND NESTED CUSTOMARY SEA-TENURE (CST) INSTITUTIONS

A Solomon Islands case study

Shankar Aswani

Introduction

This ethnographic case study illustrates indigenous non-government associated polycentric/multilevel and nested sea-tenure governance systems in the Western Solomon Islands. I examine the role of central apex-related actors in marine governance and scale-up to examine the role of family, community, tribe, confederation, local community-based organizations (CBOs), and Church indigenous marine control. I illustrate how regional sociopolitical diversity and various historical processes have shaped a number of polycentric/multilevel institutions, and explain why understanding the pluralistic and nested nature of marine governance locally matters for designing and executing development and conservation interventions in the region more successfully. Polycentrism denotes the existence of a number of governance decision-making units that may be in a competitive and/or cooperative association with each other. Polycentrism in the commons literature has often focused on the positive attributes of these multiple nodes of authority systems in regional, national, and international contexts of resource management (Carlisle and Gruby, 2017), but less attention has been given to disaggregating *indigenous polycentric* systems at the local scale.

In the case of traditional (but dynamic) long-standing indigenous institutions, such as customary sea-tenure (CST) governance, control and concomitant decision-making regarding use and access of marine resources extends across multiple levels of indigenous authority (e.g., Gruby and Basurto, 2013), which stretch from central apex-related actors to charismatic and nativist churches. That is, local powers to establish boundaries, exclude interlopers, control fishing activities, and articulate conflict-resolution procedures when present (Bromley, 1992) exist across various nodes of power that may act independently but which are more commonly interacting with each other (Ostrom, 1990). As mentioned, in this chapter I describe a polycentric and nested system of local CST governance (beyond provincial and national governance jurisdictions) in the Western Solomon Islands, particularly the Roviana and Vonavona Lagoons, and illustrate how these systems have emerged from a complex sociopolitical history. Ultimately, understanding the complexity of local marine governance polycentrism—and not assuming their lesser complexity (which is reasonable to assume) in parity with larger-scale and more complex systems (e.g., Berkes, 2006; Carlisle and Gruby, 2017)—requires a meticulous historical and ethnographic analysis of each nested layer of local governance and an understanding of their interdependence (or not). This knowledge is crucial for nongovernmental organizations (NGOs) and development agencies in their attempt to implement successful conservation and economic development projects respectively, as different levels of indigenous governance may have diverse developmental objectives.

Ignoring this local complexity molded by a complex socioeconomic and political history—that is, *history matters*—is a recipe for village or community-based management/development failure, particularly in countries with complex social economies as the Solomon Islands (e.g., Abernethy et al., 2014; Aswani et al., 2017). Also, understanding cross-level local complexity and inherent discrepancies/contradictions is important for scaling-up management and for understanding cross-scale dynamics (Cash et al., 2006). In the literature on governance of the commons and polycentrism much attention has been given to design principles of governance systems, including the demarcation of boundaries, control of interloping, monitoring capacity, and the deployment of conflict-resolution mechanisms (when present) (Ostrom, 1990). Researchers have examined how these principles are nested across different sociocultural and political institutional levels (that may or may not be interlinked), stretching across local, regional, national, and global scales (e.g., Marshall, 2007; Vousden, 2016). While the importance of the 'indigenous' and 'local' in a polycentric system of regional and national governance has been recognized (e.g., Cousins, 2008; Gelcich, 2014; Shivakumar, 2003), most analysis refers to central state governance or co-managed systems, and the nature of the micro-nested levels of power within a locality or 'community' are rarely disaggregated into its composite parts, barring a few limited accounts (e.g., Abernethy et al., 2014).

Nested complexity also occurs at the local scale in any community in which multi cross-level governance nodes are nested within the indigenous social-economy and governance systems. Understanding this fined-grained complexity is significant because polycentric governance has been often considered to have a number of attributes that result in sustainable resource management (Bixler, 2014; Lebel et al., 2006; Marshall, 2009). Yet in the acknowledgment of scale diversity (Carlisle and Gruby, 2017) for better governance the *local* scale is often poorly understood and rarely disaggregated. This is surprising because often it is local dynamics that result in the failure of community-based resource management programs. Actually, when one accounts for the polycentric plurality of sociopolitical and customary management nodes of power within any local community, particularly in Melanesia at large, then a formalized vision of a 'community' for resource management becomes increasingly dubious (Agrawal and Gibson, 1999; Nunan, 2006). Communities, particularly in the Solomon Islands in which most land and sea estates are communally owned, have people of different tribal groups living together who maintain parallel and/or hybrid identities and possess distinct political and territorial privileges and concomitant claims. This translates into a difficult context for marine resource management and sustainability in the region. Thus, attempting to understand polycentrism without scaling down into the micro-nodes of power within lower-scale levels (e.g., the 'community') is unsuitable for attaining a fuller picture, and raises questions about the often-optimistic representation of 'polycentrism' as a good formula for natural resource use sustainability governance. A detailed ethnographic account of *indigenous polycentricism* can provide a more nuanced understanding of local-scale processes, which on average are harder to uncover than those of more institutionalized formal structures such as provincial and national governing bodies. This local understanding is particularly necessary for any attempts at community-based land and marine resource management or economic development.

The regional context

The Roviana and Vonavona Lagoons are adjacent to the Island of New Georgia in the Western Solomon Islands (Figure 7.1). Land and sea estates belong to diverse tribal groups and are controlled by chiefs and elders. In terms of the sea, larger marine territorial units are often a collective of smaller tribal estates, particularly in the Eastern Roviana Lagoon (Kalikoqu and Saikile), whereas in the Western Roviana area and parts of Vonavona (e.g., Dunde and Kindu) they tend to be independent chiefdoms with a smaller demarcated marine territorial unit (Figure 7.2). Land ownership is somewhat different than ocean tenure because land modifications allow for individuals, families and smaller tribal units to carve pieces of land away from larger tribal estates into *de facto* private land units—private insofar as smaller groups of people hold the use and guardianship of the land in usufruct or perpetuity. Unlike land which can be segregated into smaller units through its physical modifications—as planting with permission of chiefs and elders can confer usufruct and/or permanent rights to a person and/or family—reefs are in principle collectively owned by the larger tribal units and guarded by tribal chiefs and elders. This, however, is not set in stone and often people will protect/manage reefs adjacent to their land shares (*hinia*), albeit attempts to close access to larger reef areas by families or smaller tribal units will be ignored by the collective unless chiefs and elders of the tribal estate (e.g., Kalikoqu) agree to temporally close an area to restock it for an upcoming social event. This system, however, it is in itself dynamic and always changing depending on the social, economic, and political currents of the time. In fact, CST rights and concomitant execution of sea entitlements have contracted and expanded into monocentric (absolute chiefly control) and polycentric (variegated control) management respectively depending on socioeconomic and political

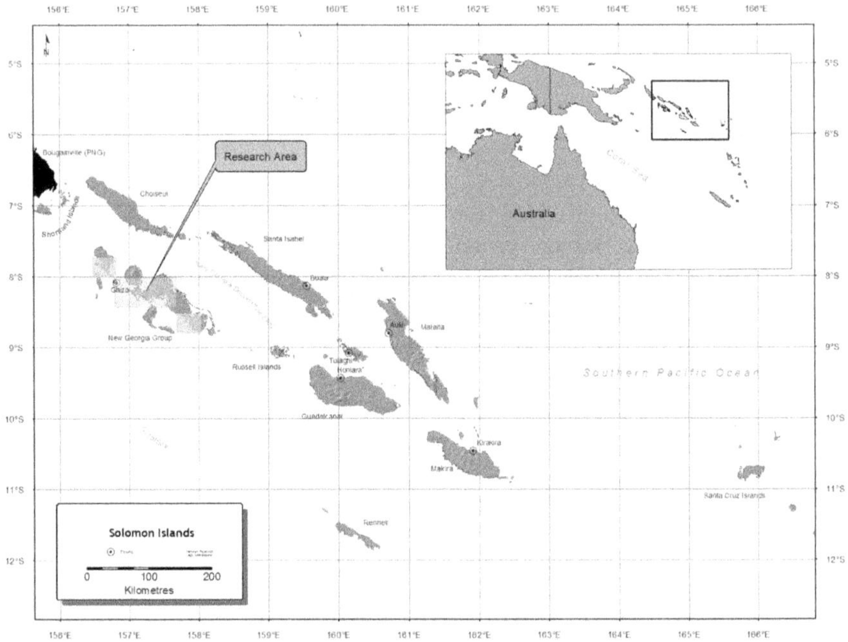

FIGURE 7.1 Solomon Islands. Research area in gray (Roviana and Vonavona Lagoons).

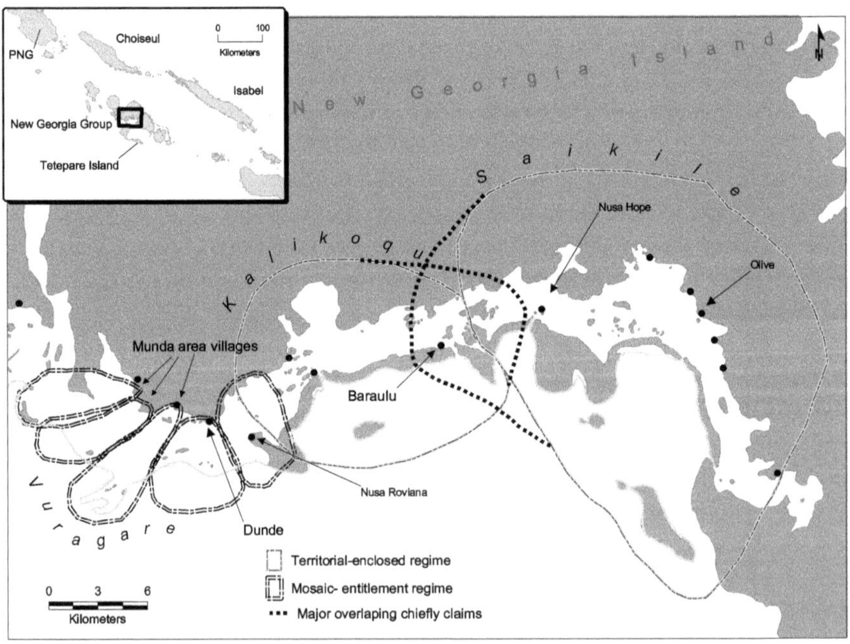

FIGURE 7.2 Roviana and Munda Areas illustrating two territorial regimes.
Source: Aswani (2005).

circumstances for at least the last two hundred years. Individuals, families, tribes, or tribal confederations (or any combination of these) may attempt to restrict access to the ocean if: (1) the social, economic, and political benefits of defending a territory offset the costs; (2) people can have their territorial claims recognized by comprising and/or adjacent groups (who may also be owners); and (3) people's tangible capacity to enforce their claims (Aswani et al., 2017).

The origin of indigenous polycentric governance systems of CST in the Western Solomon Islands emanates from a complex sociopolitical and economic history that stretches centuries back into the past. Prior to Western contact (from around the mid-19th century for sustained contact), sociopolitical stratification and associated tenurial rights and nested strengths fell into a number of designations in the Roviana and Vonavona Lagoons (Figure 7.2). These included *nabulu* (servant), *pinausu* (adopted child/servant), *tie kumakumana* (common folk), *varane* (warrior), *hiama* (priests, which was broken into two types) and *bangara* (chief, which was further broken into three sub-categories). Key to this discussion is that each of these positions had individual and collective nested rights to land and sea areas and associated resources. The strength of these rights not only was dependent on one's position but more importantly on one's kinship and association to apical land-owning ancestors. The lagoons' kinship system is generally understood as cognatic, whereby one traces their descent through cumulative links to paternal and maternal ancestors, who through initial occupation, tribal warfare, and/or through indigenous exchange occupied a given territory and constructed ancestor shrines as a recognition to his/her ancestors and as a signifier of land and sea tenure (see Aswani, 1999). People can claim rights to traditional estates either by matrilineal and/or patrilineal descent and accrue these rights cumulatively if they are associated with various land-owning ancestors across their bilinear descent line. This situation becomes more complex across geographical areas because these processes are not regionally symmetric. That is, in the Eastern Roviana area there is a cumulative matrifiliation bias (CST links are accrued via female descent), whereas in the central (Munda area) and Western (Vonavona Lagoon) regions, cumulative descent has a patrilineal bias (albeit not everyone agrees to this) (see Hviding (1996) for a related account in the neighboring Marovo Lagoon). This creates a very complex mosaic of how land and sea entitlements are conceptualized within this particular cultural region. This complexity was exacerbated by the infiltration of Western colonial powers in the 19th century, when such entanglements led to profound local sociocultural, political, and economic transformations (see Bennett (1987) for a comprehensive review of colonial interventions in the Solomon Islands).

Today, this complex social history, concurrently with changes driven by the activities of predatory Asian logging/fishing companies and evangelistic churches, and the widespread influence of social media and internet access, has resulted in the profound transformation of the indigenous social economy. In terms of indigenous ocean governance or CST, changes have also been animated by multidimensional local sociocultural and economic developments, as set out above. This, as I have described elsewhere (Aswani, 1999), has resulted in a variety of local CST systems, or three different conceptual regimes of marine governance (territorial—enclosed regime [stakeholders are nucleated within their territory], mosaic—entitlement regime [many stakeholders are dispersed away from their territory], and transitory—estates regime [a process of mixing of the other two models]) (Figure 7.2), which as of 2019 continue to transform and erode as recognized by local stakeholders. To summarize, via local migration and intermarriage (also watering-down their rights by marrying Solomon Islanders from other regions), entitlement holders have scattered away from their ancestral estates, or, in some situations, these same processes have led to their nucleation in and/or approximating their inclusive estates—for which people have rights to through the mentioned cognatic kinship system. These local movements of people have led to asymmetric territorial entitlements, that is, affording some stakeholders more opportunities to access and control various CST estates while negating others the same equivalent possibilities.

Cross-level local complexity

The summarized processes are animated by an indigenous polycentric and nested land/sea governance complexity in which nodes of territorial power are centered in *central apex-related actors, individual and extended families, communities, tribes, tribal confederations, local community-based organizations* (CBOs), and *churches* who exercise confluent and competing control over land and marine resources. The sea, nevertheless, has been less central in ongoing struggles of power over natural resources than land. In sum, it is important to recognize this cross-level local complexity and intrinsic discrepancies and contradictions if territorial management/conservation and development initiatives are to be established in the region—which many practitioners don't and, as a result, many initiatives end up as conservation and developmental failures. I provide direct excerpts from local informants to enrich the account and provide an indigenous understanding of the reviewed processes. This information was elicited through open-ended interviews with key informant for a period expanding more than 25 years.

Central apex-related actors

Central Apex-related actors are those individuals that have particular combinations of rights affording them stronger claims to particular land and sea spaces. Individuals can belong to multiple *butubutu*, or kindred groups that share consanguine ties and who have entitlement to certain portions of land/sea (the use of this term can be scaled-up from extended families to entire tribal units (Aswani, 1999; Hviding, 1996)). Sometimes people have particular combinations of rights as a result of their parents' descent lines to various *butubutu* that gives them greater power over particular areas than their other relatives. In the Eastern Roviana Lagoon (Kalikoqu and Saikile Chieftainships) (Figure 7.2), for instance, prominent individuals can trace their ancestry through a series of females which connect them to a *butubutu*. So if one is the product of five generations of say prominent Koloi (a barrier island tribe) women (cumulative matrifiliation), he or she may, in a dispute, claim greater authority over the granting of development rights to a particular area than other closely related individuals who instead may be related to three Koloi women only. Usually, central apex-related actors are chiefs *(bangara)* or men of prominence *(palabatu)*, albeit this is not always necessarily the case. While these actors are nested within extended families and tribes, they can deploy their power individually in opposition to the desire of close relatives and/or tribes. Their claim, nonetheless, can only be validated if his or her stronger rights are recognized by others, or through the courts (local and beyond) if legally disputed. A case in point is this subjective account of Parara Island and adjacent reef ownership dispute:

> At Parara there were no people in the past, I heard. But this island was adjacent to Vuragare tribe. The Vuragare is a place facing the ocean at Nusa Roviana. The reefs are owned by Vuragare following Parara in Vonavona and all the way to Nusa Aqana near Ghizo. These are the reefs and Islands of the people of Vuragare. The paramount chief of this Island was Gona Qeto, it was his Island. The power of this Island was at the hands of a Kaleqe hope [sacred old woman], but we are not supposed to name her. The people of Vuragare are also linked to this Kaleqe Hope. Because of his bloodline and relationship to Kaleqe hope, the ownership of Parara Island was rested on the hands Mr. Gona Qeto. But Gona Qeto wanted money so he wanted to sell this Island to some white traders. But some relatives who also had rights, the chiefs of Munda, Gumi and Gemu claimed that they also had ownership of Parara and that Gona Qeto could not sell the Island.

Similarly, another informant:

> At the beginning of this century, Gona, Chief of the Vuragare tribe sold the island of Parara to a British company, Levers Bros. Pacific. While there where multiple claims to this island, Gona felt that as chief of Vuragare, the power of decision over the island was vested in him. This move however angered many claimants which saw Gona's move as a threat to their joined ownership of the island. Furthermore Reverend Goldie, saw the sale of the island as a big step towards the alienation of Roviana people. Thus in preventing Levers from buying the island of Parara, Rev. Goldie gave his advice and support to Gemu, Gumi and Lampeza in the court battle [Circa 1907] that was to reverse the sale of the island back to indigenous hands. To Gona's dismay he lost the case and the jurisdictional power of the island was vested on Gumi, Gemu and Lampeza. Their descendants are the occupants of Parara Island today.

These accounts from local informants illustrate an example of the tensions between individual and collective ownership of land and sea space. While many estates are held by common groups tracing their descent to common ancestors, it is often individual actors of prominence, who on their own representation, articulate decision making to designate who can own and/or exploit the physical environment. Their claims and decisions are sometimes successful and sometimes they are not, as illustrated by the above narratives.

Individual and extended families

Individual and extended families can have various nested members who are either central apex-related actors or have substantial rights over sections of coastal land and/or small islands. Historical and recent accounts suggest that periodically, owning families will implement a closure on resource extraction in reefs adjacent to their land and enforce such a closure. These kinds of closures are more common in islands which are owed by particular families exclusively (individual or extended). It is often the case that these islands will be jointly owned by various siblings (including first and even second cousins) or each sibling will get a portion of the island as an individual plot *(hinia)*, or both arrangements may exist concurrently. Because of the outlined fluidity of the kinship system, however, family owned estates can be disputed by blood kinsfolk who may likewise have rights to the island or even be contested by larger tribal units who may disagree with an extended family's claims. Challenging groups will usually lay inactive when the island is used for habitation and subsistence, but any commercial activity (e.g., establishing a tourist lodge) will generate the emergence of latent claims by close kinsfolk. For instance, a local 1994 account by an L.O.B (tribal acronym) tribal member within Baraulu Village, in the chieftainship of Kalikoqu, reflected on the outlined process:

> Some generations ago, three grandsons of Tagosaghe split their land and sea into three realms. Toreke got the land, Goqe got the rivers and Torekana got the reefs and islands. Since these three siblings depended on each other to sustain themselves, supervision of the land and reefs were entrusted on chiefs and elders. So the descendants of Torekana are the rightful owners of the reefs and not the chiefs as presently claimed by most. However to avoid any kind of conflicts descendants of Torekana have entrusted chiefs with the care of reefs. But we descendants of Torekana also have ownership of Gulasa Island and we have the right to close fishing grounds and shell-collecting areas to members of other families who are not L.O.B.

Twenty years later in 2014, various member of the L.O.B extended family (which forms a small sub-tribal group) decided to close a section of Gulasa Island to fishing and shell collection to other Baraulu Village families and to additional neighboring groups. The motivation for the closure was to exclude other kinsfolk not belonging to L.O.B from harvesting resources from Gulasa, but more importantly, to reinforce family/sub-tribe ownership claims over the island. However, this closure was disregarded by many Baraulu villagers who did not accept the exclusive rights claimed by L.O.B and who argued that due to the confluence of various tribal groups in the past that they also had rights to the areas claimed by various L.O.B families.

Communities (villages)

In the Roviana and Vonavona region communities are generally a group of related families, associated lineages and/or assemblages of different lineage/tribal groups living in a recognized village. While village members may control a specified land and sea territory as a collective of people, villages are often made up of people of diverse tribal groups who maintain parallel and/or hybrid tribal identities. Consequently, they may possess different political and territorial claims within and beyond their particular community. In addition, neighboring villages may claim reef adjacent to a community as theirs and not recognize their exclusive use, which can make identifying bounded marine governance systems in the region a difficult task. For instance, a Kalikoqu member explains inter- and intra-village conflict over marine resources in the late 1980s and early 1990s:

> In the last few years there has been an increasing animosity among the residents of Nusa Roviana. While everyone belongs to the Kokorapa *butubutu*, some members have ties to the Vuragare tribe while others to the Kazukuru tribe [actually all those claiming to be Vuragare are also Kazukuru, but not the inverse]. As Taiyo [a Japanese company that began operating in the Roviana and Vonavona region in the early 1970s] began making payments to different neighboring groups (Kalikoqu, Dunde, Kindu, etc.) claiming Vuragare rights over the reefs of Nusa Roviana, the Nusa Roviana community drifted apart. Those having some form of affiliation to the Vuragare tribe began to call themselves Kokorapa-Vuragare, while those having no affiliation to the Vuragare tribe called themselves Kokorapa- Kazukuru. The Kokorapa-Vuragare section claimed that all reefs to the West (Vuragare) belonged to them and kin in neighboring villages [who were already collecting bait fishing fees] thus they had the right to receive the royalty payments paid by Taiyo. On the other hand the Kokorapa-Kazukuru section of the village argued that the islands to the West [close to the Munda Bar] and their adjacent reefs had been ceased to the Kokorapa-Kazukuru tribe when they paid kokopa [a form of shell money compensation] to the Vuragare tribe. This animosity festered and eventually all those involved found a compromise to avoid further disputes and the possibility of violence.

In this example, neighboring villages were claiming rights over the reefs of Nusa Roviana (and the associated fishing royalties) through their association to the 'Vuragare Tribe.' Simultaneously, members of the Nusa Roviana community who had differential rights began disaggregating their tribal identity and associated Vuragare CST rights from those of their community who did not. In this case, inter- and intra-community competition over CST rights resulted in neighboring villages as well as members of the community with differential rights to reap the benefits of the development initiative. Hence, it is important not to reify communities and villages as homogenous and cooperating units, as these are clearly not in this region, and dissention in seemingly harmonious communities can spark at the introduction of monetary benefits.

Tribes

The term 'tribe' or *butubutu* in Roviana denotes a number of units of social organization that can be scaled-up to encompass from extended families to whole regional tribal confederations. Belonging to a tribe entails that stakeholders can show their genealogical links to a founding ancestor, but how deep one delves into history will denote the breadth of a tribal group. For instance, if someone in the Kalikoqu Chieftainship used 'Votazi' (denoting three siblings going back to the 1940s) as his reference for tribal identity, this would confer him or her tribal right to Baraulu village and proximity (as well as to other areas due to cumulative entitlements through intermarriage). If that same individual mentioned 'Gutu [Qutu]' as his or her ancestor (from whom Votazi also descend), then that would give him or her rights to the use and access off terrestrial and marine resources in the Kalikoqu Chieftainship (which is made up of various intermarried tribes). However, if that same individual went back to 'Roviana' as the apical ancestor, this wouldn't mean much, as most people in the region can trace back their ancestry to this individual (some dispute its existence as a person). So when we speak about a 'tribe' in this context it entails groups of people that trace their genealogy to proximate ancestors (say, 2–4 generations), which tend to have demarcated terrestrial and marine territories and in which the strength of one's entitlement depends on the kin proximity or distance to the core tribal ancestor. These tribes tend to be nested within larger political units made up of several tribes, which through widespread intermarriage lead to the formation of larger polities. An informant explains a version of the origin of Baraulu Village (and neighboring Bulelavata as an eventual splinter of Baraulu) as an example of how movements of extended families can give raise to smaller tribes:

> Naule and Bikala moved there [Baraulu Village]. The people who lived at Baraulu in the past who built their hope [shrines], where killed all the way to Saikile, or the Koloi people. Only those living near Odikana [Chief of Saikile] were not killed. Not only the Kazukuru people killed those living in the barrier islands of Roviana [in reference to Koloi people], but killed people from Bilua, Ranongga, Duke, Lokuru, and Bangiata. The people of Koloi lived there before [Baraulu], although some of them may be Koloi,

> I am not sure because in the past these were exterminated. These stories are not too clear, they have to be straightened. In the past these were killed. Naule and Bikala were brothers. Bikala gave birth to Keto and Keto to me [the informant] and Benjamin [his brother]. It was Vagimata [Chief of Saikile] who authorized Bikala to go and stay at Baraulu. The people of Koloi were killed and those who survived stayed in the bush like wild people [Koloi descendants counter claim the land and nearby reefs of Baraulu], some stayed in some settlement. Bikala and Naule who were under the care of Vagimata farmed and fished, and established the village.

Movements of families or smaller tribes were often propitiated to access land and sea resources and/or warfare related resettlement movement. This has had a cumulative effect on the variegated nature of land and sea-tenure claims across the Roviana and Vonavona Lagoons.

Tribal confederations

As discussed above, tribal confederations are political formations that nest tribes and sub-tribes within to form larger political and territorial polities, which control extensive terrestrial and marine territories. As explained, these larger confederations nest the tenure rights of tribes, communities, families and individuals to particular composite land and sea estates. In general these confederations (particularly Kalikoqu and Saikile in the Eastern Roviana region) are different tribal groups, which share a linguistic and cultural distinctiveness and which have pooled their land and sea estates (often through intermarriage) under the control of a paramount chief. Note that chiefs are assigned 'guardianship' *(kinopu)* and not 'ownership' *(tinaqo)* over the land and sea of these large estates, albeit they still possess their own areas. While large development decisions can take place effectively at the confederation level (e.g., granting access to logging companies), tribal confederations are exposed to continuous contestation over tribal land and sea by the previously reviewed nested and polycentric units of traditional governance. An informant explains political schism as a founding process of tribal confederation formation:

> The time of Tae Bangara is when people moved to Saikile [7–9 generations ago] and found a way to have him murdered. Tae Bangara was a chief of Kalikoqu and had a wife at Kalikoqu and one at Matakota. Tae Bangara was a violent person and killed people in his area. People say that Ididu Bangara was the only chief, but there were 7 others, although he was a big chief who controlled the coast and bush. You see, the chiefly lines coming from the chief began to split and move elsewhere in parallel lines, and that is why nobody has a clear idea of the direct descent and central power of a Roviana chief. It was when the white men came and chose Ingava as the powerful chief, chief of all people. But there were many chiefs who had supporting tribes and different lands. The chiefs of Munda could claim

> links to Ididu Bangara. In the Saikile case it was Agagikana and Odikana that moved to Saikile, after the latter ordered the killing of Tae Bangara. The people who came to keep the chief of Saikile, Odikana, were the three tribes of Sosopo, Vasilavata, and Dugaha. Also, the people of Kalena Bay and their association to Saikile are older in the distant past before but they intermarried and came under Saikile. Vasilavata was the chiefly tribe and Dugaha and Sosopo were the warrior tribes. Each had their own land and islands, but these were now kept under the umbrella of the Saikile chief.

Political splintering was often triggered by indigenous warfare over territorial control. In the case of the Saikile polity, various groups associated with the chieftainship through intermarriage pooled their land and sea estates together to form a larger polity, albeit the composite tribal and sub-tribal appellations persist to this day (Saikile residents today belong to the three tribes and to other sub-tribes within) as well as their territorial claims. These kinds of processes of political expansion and contraction are widespread across the whole region, and lead to the formation of different political and territorial units.

Community-based organizations

In the Roviana and Vonavona lagoons there is only one community-based organization that pools all the confederations and lesser chieftainships under one umbrella. The CBO, which was created by myself and Roviana chiefs and elders, is called the Roviana Conservation Foundation (RCF) and this cross-level node of power was established for marine resource management and conservation. More specifically, RCF's main objective was to create networks of marine reserves for safeguarding marine ecosystems in a culturally appropriate fashion (including local governance and knowledge). RCF is formed by various regional MPA networks (Kalikile, Buki and Kiko) made-up of Resource Management Committees (RMCs), which represent the partaking hamlets within the regional territorial estates, and each RMC is composed of various village stakeholders. The responsibilities of the CBO leadership are to work with each RMC and monitor and protect the marine reserves. RCF has an elected directorship composed of prominent regional men and women, which meets annually to discuss marine protection matters. From the late 90s through around 2014 many of these MPAs were relatively successful. However, lack of sustained funding, shifting of personnel, leadership corruption and disputes related to the increased presence of logging companies obliterated the role and presence of RCF in the lagoons. As of 2019, almost all the MPAs (23 in the Roviana and Vonavona lagoons) had been disbanded and only a few villages are observing marginally some kind of resource management practices. In sum, because CST governance systems are highly contextual and contested, and there exists among regional stakeholders' different abilities to establish effective resource management, the CBO eventually collapsed as the various polycentric units described in this paper pulled its interests in different directions. In fact, the involvement of foreign universities and donors also prompted many disputes over the control of the CBO's monetary and capital assets—leading to its demise. There is a regional interest to revive the CBO and reestablish the MPAs but there is a dissonance between the different polycentric nested nodes of power to create a common front to protect the marine environment. The resurgence of the MPAs driven by local initiatives would be a good option as an adaptation strategy to the increasing negative effects of climate change and marine resource degradation in the region.

Churches

In the Roviana and Vonavona Lagoons, Christianity has been around since the early 20th century. In regards to overarching influence in this region, the Christian Fellowship Church (CFC) stands as an illustrative example. The CFC was established in the late 1950s as a breakaway from the Methodist church, combining Christian and indigenous beliefs and focusing on autonomous rule and rural development. There are over ten thousand followers spread across the Western Solomons, with the Roviana and Vonavona Lagoons having the greatest number of CFC followers. In respect to land ownership and use, while individual tribal groups and confederations have customary tenure over their estates, the CFC wields great power as an arbitrator in land/sea exploitation issues and in determining the kinds of economic development that follow. The CFC exerts influence over all the other polycentric nodes of power because it is an indigenous church, i.e., it is controlled by local peoples themselves, and because it professes collectivism in the name of self-reliance and independence. An important CFC church leader explains,

> The notion of standalone *(turu eke)* is very important to us. If you took money out of the pockets of CFC faithful, then you would weaken them and weaken their effort/work. We work in projects to fund the church, fund the tribes, and to fund the government …you see, and that is the center of the church, self relienace….so our sacrifice of CFC is to earn a decent living and to help itself, so the sacrifice of CFC is to help itself. Also with our work we can help other churches and tribes in need as well as pay taxes and assist the government. From the beginning with our founder Holy Mama, the idea was that in the coming of civilization…to create indigenism and the bringing of civilization, but in terms of indigenous peoples. You see the government today, the corruption comes from the influence of outside forces, but our principle is that if we did it ourselves within the indigenous context, we could transcend the problems created by outside influences. So that is the position of the CFC leadership and the transfer into the new generation that we still hold today…We strive to develop for self-reliance. In this process some villages are slower or faster than others. You see we bring all these young men from other villages to work here, they experience the value of work, the effort, and they will carry this to their communities. Here the development is fast, the point is

> this: the example given by the holy leadership is so that we can follow... in the context of the third movement (creation) [which began in 2014], the young men that are here, the new generation, with time and experience they will help their communities, which we already have projects in many communities, but it will take time, but they will go back and set example. You see, as Kuri [Holy Mama's brother] told us, the capital of the CFC, or the non-cash capital of the CFC, is: peace unity and love, things of the heart. If communities possess these attributes they will be strong, if not they will be weak. So when you talk about these you talk about individual righteousness, so some will be strong and some will be weak. CFC people are not angels, we are people, and we follow our Holy leaders who set the examples.

Because of the CFC influence on matters of development across its communities, it exercised great influence concerning the MPA network described in the previous section. In the early 2000s the CFC allowed and recognized the MPA network managed by the mentioned CBO (RCF) in tandem with other MPAs established in the Marovo Lagoon. For the church leadership, 'conservation' was understood as a means for enhancing local fisheries and not biodiversity conservation, and we presented the initiative as another kind of 'development' for helping rural folk in matters of food security. The CFC endorsed MPA system worked well until around 2014 when things began to unravel. This confluence between the church, the CBO, confederations and tribal groups and myself was sustained by the spiritual leader (Ikan Rove, a.k.a Papa Hope) of the church who was a champion for the MPA network. However, as his health weakened, disagreements between different stakeholders developed as the various nodes of polycentric power under the church began to raise questions about foreign interventions on church matters. Furthermore, as a consequence of its ailing leader, the CFC was split into groups A and B, and hostility between group members deteriorated after the spiritual leader's death in 2014. This hostility carried over the conservation initiatives and by 2016 most of the MPAs had been dismantled, particularly because various families and tribes demonstrated their tenurial independence and power through dissention. The vulnerability of a program driven by the charisma of the CFC church leader, or the dependence on a 'champion,' became evident when the leader passed away. The experience demonstrated local churches can have great moral authority and capacity to mobilize its flock to encourage development and conservation, but also revealed the fragility of the system if all polycentric nodes of power are not carefully synchronized to work as one—a very challenging if not impossible achievement.

Discussion

This chapter has illustrated the nested and polycentric nature of marine governance systems in the Roviana and Vonavona lagoons. It also demonstrates that nested cross-level complexity also exists locally whereby communities themselves nest polycentric nodes of marine governance that create tensions within the indigenous social-economy and governance arrangements. Essentially, CST in Roviana and Vonavona, as in the rest of the Solomon Islands, is a system of indigenous polycentric rights that nest individual, family, community, tribe, confederation, CBOs and Church levels of tenurial power. These forces may work together and/or against each other contingent on the social and economic context of the moment. An important lesson from this discussion is that multiple scales of land and marine governance have resulted from various historical processes, and that ignoring these is a formula for unsuccessful conservation or development initiatives, particularly at this junction in history when the Roviana and Vonavona regions are being opened to tourism development and people are increasingly accessing digital networks. Conservation and development practitioners need to cognize that the indigenous tenure systems that govern resource use and access rights are functioning at various scales that are contextual in space and time.

Specialists working in the region need to assess how, when, and where the reviewed levels of governance are in action and in partnership with local people address the concerns of each nested level. They also need to recognize that development and resource management efforts are likely to expand and contract over time because of the likelihood that the various polycentric nodes of power desynchronize to oppose and face each other. Government officials and funding agencies need to offer development and conservation practitioners the means to ascertain the intricacies of how these nested and polycentric governance systems function in order to inform place-based conservation and development. Ignoring these local intricacies will only lead to failure, as decades of rural development and conservation initiatives in Melanesia have shown. In sum, the ethnographic analysis of these local nested and polycentric systems provides a way to scale-down and examine communities more carefully, and to uncover complexities and discrepancies that an essentialized view of the 'local' or 'community' does not provide (see also Agrawal and Gibson, 1999; Nunan, 2006). It also challenges the notion, at least at the local scale, that polycentric governance is more likely to result in sustainable resource management than not. This ethnographic account provides a more nuanced analysis of cross-level dynamics by discerning the kinds of local decision-making authority that different actors and/or groups possess, how they interact, and the various nested layers of traditional entitlements shaped by a complex socioeconomic and political history. This knowledge is fundamental to formulate a cross-scale analysis and to design successful marine conservation and development programs in Melanesia and beyond.

References

Abernethy, K. E., Bodin, Ö., Olsson, P., Hilly, Z. and Schwarz, A. (2014) 'Two steps forward, two steps back: The role of innovation in transforming towards community-based marine resource management in Solomon Islands', *Global Environmental Change*, vol. 28, pp. 309–321.

Agrawal, A. and Gibson, C. C. (1999) 'Enchantment and disenchantment: The role of community in natural resource conservation', *World Development*, vol. 27, pp. 629–649.

Aswani, S. (1999) 'Common property models of sea tenure: A case study from the Roviana and Vonavona Lagoons, New Georgia, Solomon Islands', *Human Ecology*, vol. 27, no. 3, pp. 417–453.

Aswani, S. (2005) 'Customary sea tenure in oceania as a case of rights-based fishery management: Does it work?' *Reviews in Fish Biology and Fisheries*, vol. 15, no. 3, pp. 285–307.

Aswani, S., Albert, S. and Love, M. (2017) 'One size does not fit all: Critical insights for effective community-based resource management in Melanesia', *Marine Policy*, vol. 81, pp. 381–391.

Bennett, J. A. (1987) *Wealth of the Solomons: A History of a Pacific Archipelago, 1800–1978*. University of Hawaii Press, Honolulu, Hawaii.

Berkes, F. (2006) 'From community-based resource management to complex systems: The scale issue and marine commons', *Ecology and Society*, vol. 11, no. 1, p. 45.

Bixler, R. P. (2014) 'From community forest management to polycentric governance: Assessing evidence from the bottom up', *Society & Natural Resources*, vol. 27, no. 2, pp. 155–169.

Bromley, D. W. (1992) 'The commons, common property, and environmental policy', *Environmental and Resource Economics*, vol. 2, no. 1, pp. 1–17.

Carlisle, K. and Gruby, R. L. (2017) 'Polycentric systems of governance: A theoretical model for the commons', *Policy Studies Journal*, early view. Online: https://onlinelibrary.wiley.com/doi/full/10.1111/psj.12212

Cash, D. W., Adger, W., Berkes, F., Garden, P., Lebel, L., Olsson, P., Pritchard, L. and Young, O. (2006) 'Scale and cross-scale dynamics: Governance and information in a multilevel world', *Ecology and Society*, vol. 11, no. 2, p. 8.

Cousins, B. (2008) 'Characterising 'Communal' Tenure: Nested systems and flexible boundaries', in A. Claassens and B. Cousins (eds) *Land, Power and Custom: Controversies Generated by South Africa's Communal Land Rights Act*. Juta and Company Ltd, Cape Town, South Africa.

Gelcich, S. (2014) 'Towards polycentric governance of small-scale fisheries: Insights from the new 'Management Plans' policy in Chile', *Aquatic Conservation: Marine and Freshwater Ecosystems*, vol. 24, no. 5, pp. 575–581.

Gruby, R. L. and Basurto, X. (2013) 'Multi-level governance for large marine commons: Politics and polycentricity in Palau's protected area network', *Environmental Science and Policy*, vol. 33, pp. 260–272.

Hviding, E. (1996) *Guardians of Marovo Lagoon: Practice, Place, and Politics in Maritime Melanesia* (Vol. 14). University of Hawaii Press, Hawaii.

Lebel, L., Anderies, J. M., Campbell, B., Folke, C., Hatfield-Dodds, S., Hughes, T. P. and Wilson, J. (2006) 'Governance and the capacity to manage resilience in regional social-ecological systems', *Ecology and Society*, vol. 11, no. 1, p. 19.

Nunan, F. (2006) 'Empowerment and institutions: Managing fisheries in Uganda', *World Development*, vol. 34, no. 7, pp. 1316–1332.

Marshall, G. R. (2007) 'Nesting, subsidiarity, and community-based environmental governance beyond the local scale', *International Journal of the Commons*, vol. 2, no. 1, pp. 75–97.

Marshall, G. R. (2009) 'Polycentricity, reciprocity, and farmer adoption of conservation practices under community-based governance', *Ecological Economics*, vol. 68, no. 5, pp. 1507–1520.

Ostrom, E. (1990) *Governing the Commons*. Cambridge University Press, New York.

Shivakumar, S. J. (2003) 'The place of indigenous institutions in constitutional order', *Constitutional Political Economy*, vol. 14, no. 1, pp. 3–21.

Vousden, D. (2016) 'Local to regional polycentric governance approaches within the Agulhas and Somali Current Large Marine Ecosystems', *Environmental Development*, vol. 17, pp. 277–286.

8

POLITICAL ECOLOGIES OF RESOURCE GOVERNANCE

Ontologies, agency and practice

Caroline Upton

Introduction

Effective and equitable governance of renewable natural resources remains a major concern for a diverse range of global and local actors. An increasing sense of urgency occasioned by the ever-accelerating impacts of climate change and biodiversity loss further exacerbates contemporary governance challenges. Whilst a range of specific governance frameworks exist through which these challenges can be theorised and examined, as explored elsewhere in this volume, political ecology (PE) continues to offer an important and often underutilised complementary approach.

As a loose body of theory, approach or frame of enquiry, PE has been widely applied by activists and scholars across a range of sectors and geographical contexts to shed light on critical human–environment intersections and their effects. For example, PE has been variously deployed as a lens through which to critique conservation politics and the construction of 'neoliberal natures' (Sullivan, 2017); to examine resource access and practices in relation to wider political and economic structures and to highlight the knowledge politics and power involved in discursive creation of environmental crises (Fairhead and Leach, 1996). Post-structural PE in particular has been central in illuminating and challenging the positioning of local actors as drivers of resource degradation and over use in diverse contexts.

It is notable that critiques of other approaches and frameworks for natural resource governance, as presented elsewhere in this volume, highlight a lack of serious attention to power as a key weakness. PE has much to offer here, especially through recent work which develops theorisations of power therein, and links these to growing deployment of governmentality as a framing device in conservation/natural resource governance (Fletcher, 2010; Svarstad et al., 2018).

Increasing concerns in contemporary PE with diverse ontologies and the role of non-human actors therein suggests further productive engagements with governance debates (Escobar, 2010).

In the following chapter these issues are explored with particular reference to cross-sectoral governance debates and challenges in Mongolia. As a country in which the author has worked for many years, Mongolia in many ways offers a microcosm of a range of intersecting governance challenges. From its extensive rangelands, still largely de facto managed as common property regimes, and from which some 28% of the population derive predominantly pastoralist livelihoods, to emergent mining and carbon politics and economies, all located in the same rural spaces as a globally important biodiversity heritage, governance is complex, multifaceted and rapidly changing. Following an introduction to PE and the Mongolian context, this chapter goes on to explore how a PE framing is uniquely positioned to shed light on these governance dilemmas.

PE and governance

'Political ecology' as a concept continues to elude a unanimous definition. It has been variously described as 'a trans-disciplinary attempt to integrate natural and social science approaches to understanding the relationship between human and ecological systems' (Peterson, 2000: 323); as constituting '… empirical, research-based explorations to explain linkages in the condition and change of social/environmental systems, with explicit considerations of relations of power' (Robbins, 2004: 12) and as 'a concern with tracing the genealogy of narratives concerning 'the environment', with identifying power relationships supported by such narratives, and with asserting the consequences of hegemony over, and within, these narratives for economic and social development, and particularly for constraining possibilities for self-determination' (Stott and Sullivan, 2000: 2). PE is here understood as a loose body of theory, approach or frame of enquiry, targeted on the above issues, and within which relations of power are central. As argued by Nunan (2015: 30), 'at [political ecology's] heart is a concern with the nature and influence of power in influencing how environmental degradation is understood, explained and manifested'. Thus, when looking at governance through a PE lens, concerns with power relationships encompass not only tangible issues of resource access and usage but also discursive constructions of 'crisis' and competing explanations of environmental change and the role of diverse actors therein.

Implicitly positioned in contrast to an 'apolitical ecology', the early proponents of the PE approach, which emerged powerfully in the 1970s, were concerned to 'denaturalise' and challenge prevalent often neo-Malthusian inspired accounts of environmental change. In such accounts, the pressures of growing populations on a limited natural resource base formed the implicit framing, with local resource users often positioned therein as agents of environmental degradation, driven by poverty to excessive or irrational resource use in the short term, with little care for longer time horizons.

These (for some convenient) caricatures were challenged by early political ecologists in a number of ways. One was through classic 'chains of explanation' approaches by which leading proponents such as Piers Blaikie sought to place phenomena such as rural soil erosion in Nepal with their wider political and economic contexts (Blaikie and Brookfield, 1987). In such neo-Marxist accounts, the existence of soil erosion, in the Nepal case linked to local farming practices, was not disputed; rather the focus was shifted from local farmers as primarily responsible for the phenomenon, to the policies pursued by the Nepalese government which forced farmers' displacement onto marginal lands and thus underpinned observed degradation (ibid.). The more recent evolution of post-structuralist PE approaches has placed the emphasis on discourse and the politics of knowledge, or as expressed by Goldman et al. (2018: 2) with reference to climate change, 'the persistently uneven power dynamics that shape what is considered valid knowledge'. Through such post-structuralist PE approaches the very existence of particular environmental problems and crises has been called into question; not only prevalent explanations and allocations of responsibility for their occurrence. Classic studies in this vein include Fairhead and Leach's (1996) study of alleged deforestation in Kissidougou, West Africa; wherein local villagers' accounts of their creation of forest areas, rather than widely held policy discourses of locally driven deforestation, were upheld by the evidence but persistently silenced in governance debates. In governance terms these representations therefore matter profoundly, in that they underpin policy prescriptions, may be used to contest meaningful devolution of rights and power and often exert pernicious influence on local livelihoods and practices.

However, despite the potential of a PE approach to shed light on these important aspects of governance, particularly through attention to power and its expressions, recent critiques have suggested that the latter remains insufficiently theorised in PE. Notable here is the work of Svarstad et al. (2018: 351), who concur that 'power is at the core of political ecology as a scholarly approach', yet argue that conceptualisation of how it operates, necessary for more productive contributions to governance dilemmas, remains weak. Svarstad et al. (ibid.) recognise actor-oriented perspectives and neo-Marxist thinking on power in PE which variously examine agency of specific actors and its structural and historical constraints, the latter often deploying a 'chains of explanation' approach linked to global capitalism. Post-structuralist perspectives, outlined above, are further expanded therein to build on Foucault's concept of governmentality, an increasingly important way in which PE engages with the workings of power in environmental governance (Fletcher, 2010; ibid.). Famously described as 'the conduct of conduct', governmentality in PE offers new insights into the operation and effects of natural resource governance, as well as its limitations. Fletcher (2010) proposes a novel framework, based on Foucault's four forms of governmentality, namely discipline, sovereignty, neoliberalism and truth governmentality, as applied to contemporary biodiversity conservation. He argues that these intersecting approaches and 'strategies of governing', variously denoting internalisation of state-led or externally derived norms (disciplinary governmentality), top down enforcement of laws and regulations (sovereign governmentality), governance through external market incentives (neoliberal governmentality) and governance through beliefs in particular truths or ontologies about human–environment relations (truth governmentality), together offer explanatory power in conservation governance. As explored further below in relation to the Mongolia context, this chapter argues that this approach can usefully be applied to natural resource governance more broadly, as part of a multifaceted PE approach.

A further emergent strand of this multifaceted PE which merits attention here is what has been described as 'a political ecology of difference' or 'political ontology'. According to Escobar (2010), this is grounded in explicit recognition of non-Western ontologies of human–environment relations and ways of knowing nature. For example, Blaser (2013: 14) notes with reference to an ostensible struggle over resource rights in the Peruvian Amazon, that for indigenous peoples these struggles centred less on issues of access per se, but rather on concerns with 'defending complex webs of relations between humans and nonhumans, relations that, for them, are better expressed in the language of kinship than in the language of property'. Attention to political ontology thus unsettles universalisms and admits of diverse, multiple realities or socionatural orders, and ways of being in and knowing 'nature'. When applied to governance debates, this ontological turn thus offers the prospect of further challenging and decentring western knowledges, and assumed universal truths, if diverse worlds and ontologies are taken seriously.

In terms of natural resource governance dilemmas, these facets and potential contributions of PE may be usefully considered in relation to questions characteristic of a PE approach, namely, (i) how do nature and society influence each other and shape access to natural resources?; (ii) how are concepts of nature and environment constructed and with what implications for human-environment interactions?, (iii) What are the linkages between resource access/control, livelihoods and environmental change, and (iv) what are the social impacts of environmental change? (Schubert, 2005).

In summary, PE has famously been characterised as both a hatchet and a seed through 'critically explaining what is wrong with dominant accounts of environmental change, while at the same time exploring alternatives, adaptations, and creative human action in the face of mismanagement and exploitation' (Robbins, 2012: 20). Despite this characterisation, it has in the past been widely criticised for a lack of policy relevance and articulation. The emergent strands of a revitalised, contemporary PE, as sketched out above, offer the prospect of some redress, not least through PE's applications to the critical understanding of contemporary natural resource governance dilemmas. It is to these the chapter turns, following a brief introduction to the Mongolian context.

Mongolian context

Despite growing trends of urbanisation and sedentarisation, mobile or 'nomadic' pastoralism retains prominence in Mongolia's extensive rural spaces, wherein it constitutes a key livelihood strategy for some 28% of the country's

3.17 million population. It also remains integral to both individual and national identities. The governance dilemmas discussed herein apply to these vast rural landscapes, some 80% of which comprise unimproved grasslands. Renewable natural resources at stake here thus include pastures, water resources and the range of species and habitats that contribute to Mongolia's globally important biodiversity. The recent construction of carbon as a commodity through international climate politics (e.g. Kyoto's Clean Development Mechanism and more recently REDD+) has effectively added another resource to Mongolia's rural spaces.

Governance dilemmas derive from and are increasingly exacerbated by competing and diverse use of and claims to renewable natural resources, linked to the interests of local actors; to government agendas across scales; and to the growing reach of donor discourses and practices, particularly around carbon and conservation. The status of Mongolia's pasturelands as de facto commons, officially in state ownership, but with more exclusive rights of particular herding families to specific winter and spring camps, is an important factor here. Others include the burgeoning mining economy, worth more than 80% of export earnings by 2012, but inevitably impinging spatially on renewable natural resources integral to herders' livelihoods. Mongolia's declared ambitions to extend its Protected Area coverage to some 30% of the country suggest further spatial trade-offs and struggles over resource access and governance.

Key actors in the governance challenges explored herein thus comprise pastoralists, in other words those who manage and practice pastoralism on a daily basis and for whom this constitutes their main livelihood strategy. In addition to the 28% of the population officially categorised as pastoralists, their extensive social/kinship networks typically include urban-based extended family members who may be temporarily absent from herding households (e.g. for education or training). They also include those who are more permanently based in Mongolia's urban centres, typically the capital Ulaanbaatar, but who may return to 'the countryside' over the summer months each year and for whom rural relatives may care for livestock herds. Thus, multiple lives and livelihoods are entangled in the use and governance of Mongolia's pasturelands and pastoral economies, over and above the already substantial percentage suggested by official government statistics. Other key actors include those placed within the various tiers of government/administration, from central government in Ulaanbaatar to the devolved bodies at *aimag* (region), *soum* (district) and *bag* (village) level. These include local governors and specialist actors, for example land management officers, charged with particular responsibilities in relation to governance of pasture and its users.

Still within the realm of human actors, international donor/'development' agencies have had an increasingly prominent role in Mongolia since the end of the '*negdel*' (Soviet-inspired collective) era and rural decollectivisation in the early 1990s. The importance of aid flows at the time led Mongolia to be considered one of the most aid-dependent economies in the world, with up to 40% of GDP being derived therein at particular times following collapse of the Soviet

Union. In specific relation to governance of pastoral lands and livelihoods, these external interventions have variously focused on Mongolia's globally important biodiversity, through the ecosystems, habitats and specific species that co-exist with pastoralists (e.g. GTZ Nature Conservation and Bufferzone Development Project); on sustainable pastoral livelihoods and rural development (World Bank Sustainable Livelihoods Project); and more recently on climate change adaptation and resilience (MOR2 Project). Of course, these are not mutually exclusive but are entangled both in specific spaces, where donor programmes overlap and compete, and even within single programmes, for example where 'win-win' solutions between conservation and livelihoods continue to be espoused.

The emergent and now burgeoning mining economy has introduced a new range of actors into resource governance debates, as the discovery of significant gold and copper deposits in particular has triggered new struggles and contestations over rural spaces. Although minerals such as copper and gold may not be deemed renewable NR, as defined in this book, they are considered herein as their governance intersects with and influences that of a range of other renewable NR in Mongolia's rural spaces. Deriving from these encounters, environmental activists have also emerged as key actors in rural spaces, for example the Ongii River Movement in 2001 and the at one time countrywide Coalition of River Movements which evolved from this over the following decade (Upton, 2017). These activists are, or were originally, local herders/pastoralists, who may thus assume diverse and strategic identities in governance struggles. Linked both to mining-related contestations and to conservation politics, religious actors (shamans and Buddhist monks) have also figured as important governance actors in specific moments and contexts.

To complement the complexity of the human realm in Mongolia's contemporary environmental governance, a range of non-human actors also merit consideration. As observed elsewhere, these include aspects of an animist nature. High (2013: 754) emphasises that 'elected politicians are only some of the agents that appear on the political scene … with politics not necessarily delimited to the human realm'. As a core element of recently revived shamanic/Buddhist beliefs and cultural norms around nature and environment *(baigal)* is the idea that landscapes are inhabited by spiritual entities (*gazaryn ezed*, 'lords of the land'), necessitating respectful, reciprocal relations with nature (Sneath, 2014: 461). Contra to Euro-centric understandings of nature and environment, *baigal* also includes 'animals, mountains, trees, grass, weather… (etc.) … as active subjects which have their own ways of being that affect human beings, just as humans have ways of life that affect them' (Humphrey and Sneath, 1999: 3, cited in Empson, 2011: 87). Put another way, these non-human actors can be considered an integral part of a kind of 'sacred geography', wherein establishment of resource rights require favourable relations with local spiritual entities (Empson, 2011: 236).

The following sections analyse these governance challenges in relation to both human and non-human actors and through deployment of a PE lens and framework. The first and lengthiest section focuses on PEs of conservation governance, followed by shorter sections on the more recently emergent issues of carbon governance and mining. It is argued here that PE is particularly powerful in illuminating power issues and agency therein.

Political ecologies of conservation governance

In the early 1990s, decollectivisation of the rural herding economy marked Mongolia's entry into the 'Age of the Market'. As argued elsewhere (Upton, forthcoming), this opening up in political and economic terms was paralleled by the 'discovery' of Mongolia by a number of major international conservation organisations, which together with emergent global conservation regimes did much to transform the nature and scale of conservation governance and practice. For example, between 1990 and 2000 Mongolia adopted a range of key global conventions, notably the Convention on Biological Diversity (1992) and also IUCN categorisations and guidelines for Protected Areas. Its national Biodiversity Conservation Action Plan (1996) reflected these commitments and set out a new national ambition to expand its Protected Areas (PA) network to cover at least 30% of the country, from an estimated 8% at the time (MNE, 1996). Such developments were accompanied by a plethora of new conservation-oriented legislation and domestic governance institutions, as both national and international attention focused on protection of Mongolia's biodiversity heritage.

Particularly since 2000 multiple international donor organisations (e.g. World Bank, ADB, GTZ, UNDP) have supported the creation of herder groups or 'communities' in rural districts, variously designed to facilitate 'sustainable livelihoods' and/or community-based conservation, often underpinned by changing tenure regimes. According to the United Nations Development Programme (UNDP, 2007), since May 2006, donors allocated US$77.5 million to projects centred on creation and formalization of some 2000 such herders' groups, typically comprising 10–40 households. Thus, in addition to extension of the PA network, community-based natural resource governance solutions, focusing on 'win-win' solutions for livelihoods and conservation, have also extended across pastoral landscapes. These 'herder groups' continue to appear as an important governance model, ostensibly linked to devolution from central state control and enhanced local participation in resource governance (Fernandez-Gimenez et al., 2015). The 4th National Report on Conservation of Biodiversity of Mongolia confirmed government commitments to 'the participation of local people in … programmes and plans for sustainable utilization of biological diversity' within and without PAs, in accordance with community-based conservation paradigms (MNET, 2009: 61–62). However, Batbold et al. (2015: 29), contend that, despite emphasis on the sustainable use of biodiversity through development of government/citizen/private sector partnerships in current National Biodiversity strategies, both these tripartite partnerships and more traditional community-based grazing governance regimes remain challenging to enact and more evident on paper than on the ground. Reading et al. (2016) concur that limited resources and capacity constraints have negatively impacted active governance of many Mongolian Protected Areas following their establishment; a situation that persists in many areas today.

More recently, and in parallel with the conservation models outlined above, the Mongolian government has embraced the Ecosystem Services (ES) paradigm. ES-based planning and 'appropriate' valuation of biodiversity and ES has become a prominent policy framing. For example, under the current National Biodiversity Programme (2015–2025) 'taking into account of the value and importance of pasture, water resources and forest ecosystem services…' is stated as a key goal as are the development of new laws and programmes for valuing ES (Batbold et al., 2015: 34).

What then, specifically, can a PE approach or lens bring to understanding of the governance debates mapped out above? I argue that a PE approach does the following: it directs us to look at inclusion and exclusion in resource access and decision-making, through attention to issues of power and the ways these intersect with environmental change; through its post-structural turn it highlights the discursive processes that underpin issues of rights of access and decision making; and finally it illuminates diverse 'political ontologies' and politics of knowledge in governance struggles.

Analysis of resource access issues in Mongolia could and indeed often does focus on the complex, intersecting, hybrid institutions that shape and underpin access. As I have argued elsewhere, informed by aspects of common property theory and Cleaver's notion of 'institutional bricolage', herders' resource rights, particularly to pastures, are grounded to varying degrees in historical claims, forms of social organisation and direct allocation of rights through formal legislative frameworks (Upton, 2009). Institutional bricolage, adaptation and the constant reworking of rights and norms challenge static notions of customary rights, whilst a series of donor interventions around the formation of varying types of 'community groups', as outlined above, have provided a further layer of opportunity and complexity (ibid.). Viewing these through a critical PE lens, however, directs us not only to document the specific institutional frameworks that underpin resource use and access, but to go beyond this to emphasise the power relationships and discourses which produce and are reinforced by these.

For example, environmental change in Mongolia is widely positioned as a threat to biodiversity conservation. 'Overgrazing' and rangeland degradation, are also increasingly highlighted as significant barriers to the country's realisation of its Convention on Biological Diversity (CBD) and other conservation obligations. While the role of climate change and the recent proliferation of mining activities are acknowledged in these analyses of pasture degradation, the significant growth in total livestock numbers and changes in livestock composition feature prominently, with herding communities held liable for these adverse developments. Herding communities are thus discursively positioned in complex ways in relation to conservation governance. The emphasis on devolved, community-based conservation is, according to Sneath (2003), underpinned by external representations of Mongolia's pastoralists as a potential resource for conservation, uniquely receptive to externally derived and formulated conservation initiatives by virtue of their 'traditional' nomadic values. Zhang et al. (2007: 21) further argue, with specific reference to Mongolian herders, 'the core idea of nomadic culture is centred on an awe of life, respect to nature and harmonious co-existence of humans with nature ...'. Since the mid-1990s a number of Mongolian policy documents have suggested the unique potential for and desirability of integrating nomadic, 'traditional' values and practices with 'modern science and technology' in the pursuit of biodiversity conservation and sustainable development (MNE, 1996: 1; MNET, 2009; Reading et al., 2006). They are typically less clear about the contributions of specific 'traditional', cultural practices and knowledges, and about how such synergies are to be achieved. Such discursive representations may well be deemed idealised, and also sit uneasily with those of herders as drivers of environmental change and degradation, through irrational stocking and management practices. These latter power-laden representations have a long and inglorious history beyond Mongolia. Histories of conservation suggest that pastoralists globally have been especially prone to loss of land through eviction and exclusion from protected areas and the resources therein, often justified by just such representations of pastoralism as environmentally destructive and irrational (Brockington, 2002). In recent years, pastoralists' own identity politics and practices of 'strategic extraversion' have begun to push back against these though counter discourses, positioning mobile pastoralism in dryland environments as integral to realisation of global and national conservation goals (Upton, 2014).

A PE lens can, thus, as a first step, draw critical attention to these discourses, to who has the power to represent herders therein and with what effects. It highlights how identity-based pastoralists' mobilisations have begun to counter externally derived representations, through reclaiming discursive ground around the pastoralism/conservation nexus. In governance terms, a PE approach is critical in showing how actual practices and access rights are shaped by these discursive struggles. As highlighted above, policy and donor discourses position pastoralists in multiple, often conflicting ways, that both underpin ostensible moves towards community-based conservation and resource management, whilst simultaneously seeking to regulate pastoralists' 'irrational' or problematic behaviours. Mongolia's current and much debated Land Law requires local *(soum)* governors to control these behaviours by regulating numbers of livestock and stocking rates, as well as seasonal movements between pastures, though in effect they often lack the capacity to do so (Fernandez-Gimenez et al., 2015; Upton, 2012). Both capacity constraints and the principle of more participatory approaches have led to increasing emphasis on herders' involvement in this pasture use planning; something that has also been emphasised by donor projects focused on the development of 'herder groups', and which shapes pasture rights and access. For example, the Deutsche Gesellschaft fur Technische Zusammenarbeit/Mongolian Ministry of Nature and Environment (GTZ/MNE) Nature Conservation and Bufferzone

Development Project (1995–2002) and its successor 'Conservation and Sustainable Management of Natural Resources-Gobi Component' (2002–2006) project, were strongly grounded in local participation in conservation governance and, by extension, natural resource governance more widely. A particular development linked to these GTZ projects was the emergence of Community Managed Areas (CMAs), based on contracts between herder groups, local *soum* and Protected Area administrations, although the basis of these within national legislation was contested at the time (Upton, 2010). These and other donor initiatives rely to some extent on the discursive positioning of pastoralists as resources for conservation, by virtue of their 'traditional' norms, practices and institutions, often grounded in the idea that institutions such as CMAs will 'enhance herders' sense of ownership and responsibility, hence promoting conservation-oriented behaviour' (ibid., 313). The particular behaviours and access rights agreed under such CMA and comparable contracts were locally specific, although the difficult issue of stocking rates, often seen by external actors as central to 'conservation-oriented behaviour', remained problematic.

Taking this further, adopting a Foucauldian approach to power within PE, illuminates the diverse and intersecting roles of discipline, sovereignty, neoliberalism and truth governmentality. In relation to conservation issues and debates sketched out thus far, the former two modes of governmentality are most evident. As highlighted by Fletcher (2010), disciplinary governmentality seeks to shape behaviours through individuals' internalisation of particular 'desirable' norms and values. Production of herders as 'conservationists', through for example donor or state programmes in Mongolia, could thus be seen as just such a form of 'green' disciplinary governmentality, wherein what constitutes desirable or conservation-oriented behaviour may be defined by external interventions, but is intended to be internalised and acted on by herders as subjects. Such green governmentality, or environmentality, as termed by Agrawal (2005), is typically realised, for example, through forms of training, education, community workshops and capacity building, typical of 'herder group' focused interventions. Disciplinary governmentality typically operates in conjunction with other forms, in this case sovereign governmentality, whereby natural resource governance and in this case particularly conservation, is shaped by legal frameworks and the exercise of state and donor power. In the Mongolian case, the intersections of these two forms of governmentality are evident in that one may be necessary to produce or reinforce the other: the creation of 'eco herders' (a term used by some donor projects in Mongolia), for example, may be a necessary basis for the effective exercise of sovereign power in conditions of limited state capacity.

Neoliberal governmentalities are also nascent in contemporary Mongolia. Specifically, ecosystem service (ES)-based planning and 'appropriate' valuation of biodiversity and ES has become a recent priority for the Mongolian government. For example, under the current National Biodiversity Programme (2015–2025) 'taking into account the value and importance of pasture, water resources and forest ecosystem services…' is stated as a key goal, with one of the associated indicators being the 'amount of payment for ecosystem services (PES)' (Batbold et al., 2015: 34). The development of new laws and programmes for valuing ES and for implementing PES by 2024 and 2025 respectively, are stated targets under this same current National Biodiversity Programme (ibid.). A PE lens applied in this particular context would firstly highlight the concept of ES as an anthropocentric construct and seek to denaturalise it. It would point to and contest what Robertson (2012) describes as the 'remaking' of the world into a 'collection of ES'. It would also question what is made visible and what is hidden by this construct. For example, a 2009 World Bank report on the Tuul River basin is one of the few widely available ES-based evaluations in Mongolia (Emerton et al., 2009). This and the few other available examples focus almost solely on economic valuation for a limited range of provisioning ES. As I argue elsewhere (Upton, forthcoming), given the prominence afforded to PES and economic valuation of ES in contemporary conservation planning in Mongolia, it is clear that these ostensibly neoliberal approaches and frameworks will increasingly shape practice into the foreseeable future. Such 'neoliberal conservation' is at the heart of many PE-informed critiques of conservation governance (e.g. Fletcher and Buscher, 2017; Sullivan, 2017). A PE lens can and should, however, also go beyond mere critique, if it is to function not only as a hatchet but as a seed. In this case it also enables further understanding of how ostensibly neoliberal governmentality operates and how it may be subverted and resisted, in conjunction with other forms of governmentality, to produce particular practices. These issues are considered further for Mongolia in relation to intersections of conservation and climate politics in the section 'Governing Carbon', below.

Finally, as previously highlighted, non-humans may also be deemed key actors in conservation, according to particular ontologies and 'truth governmentalities', linked to pre-revolutionary and re-emergent animist (shamanic) and Buddhist beliefs. At the core of these are concepts of landscapes as inhabited by spiritual entities, which necessitate reciprocal, respectful relations with nature (Marin, 2010: 164; Sneath, 2001). In governance terms, these may take the form of cultural prohibitions on cutting live trees, digging the soil and overuse of resources, as well as participation in *ovoo* ceremonies, in pursuit of favourable environmental conditions and personal good fortune (Humphrey et al., 1993; Upton, 2010). Underpinning ontologies may thus differ from those espoused in externally derived models of community-based conservation. Arguably in Mongolia, conservation policy continues to be driven mainly by the perspectives of Western conservation biology and PA planning, with less emphasis on locally derived, diverse knowledges and ontologies, despite participatory rhetoric (King and Gurnell, 2005; Reading et al., 2006). The ontological turn in PE (Escobar, 2010) and post-structural emphases on truth governmentality illuminates these tensions and their implications for governance in practice. Again, some of these intersections are examined further in the section 'Governing Carbon' below.

To date and to the best of the author's knowledge, there are no published analyses of conservation in Mongolia that deploy a PE grounded governmentality framework. The author has one such analysis underway, with further empirically grounded details of its operation in specific rural spaces. For the purposes of this chapter, however, the point is to emphasise that it is through the application of this particular post-structuralist PE lens that power and its importance for shaping governance strategies and resultant practices become more legible.

PEs of carbon governance

In Mongolia, as in many countries across the globe, carbon has effectively been created as a resource and as an object of governance through global climate politics and resultant frameworks and agreements, for example the UN Framework Convention on Climate Change (UNFCCC) and Paris Climate Agreement. Within these, the Clean Development Mechanism (CDM) and Reducing Emissions from Deforestation and Forest Degradation (REDD+) initiatives have been central in positioning carbon as a (tradeable) resource. Mongolia, as a signatory to UNFCCC, the Kyoto Protocol and the Paris Agreement, has placed climate governance, driven by global institutions and norms, at the centre of its contemporary priorities. It is currently developing its National REDD+ strategy, focusing on forested areas, in accordance with the REDD+ programme. However, for the vast majority of the country's land area, dominated by unimproved rangelands, intersections between carbon politics, livelihoods and conservation governance are currently little developed or understood.

One exception, which illustrates the importance of a PE lens therein, is in relation to the country's first externally accredited community-based rangeland carbon project, Pastures, Conservation and Climate Action, Mongolia (PCCA). As reported in detail elsewhere (Upton, forthcoming), the PCCA Payment for Ecosystem Services (PES) scheme was set up under the Plan Vivo Standard at three pilot sites across Mongolia, with the initial phase designed to run from 2015 to 2019 inclusive. This was a collective endeavour between local herders, the Mongolian Society for Range Management (MSRM) NGO, with support from local administrations and underpinned by a preceding Darwin Initiative funded research project (2012–2015), for which the author was a co-researcher. Given the wider governmental context, through which a clear focus on ES and on PES was articulated, irrespective of any external, academic concerns with 'commodification of nature' or the extension of neoliberalism, the goal under PCCA was to create space for expression of local values and priorities. Thus, while PE critiques of PES were acknowledged, PE was also deployed to contest and examine assumptions around impacts and power relations therein. Employing Kallis et al.'s (2013) approach of 'radical pragmatism', PCCA sought to maximise opportunities for the expression and exercise of local agency and values, particularly in relation to extension of ES thinking beyond the economic sphere to incorporate 'affective socio-ecological relations' and values in any resultant ES-based interventions. In doing so, the project was attentive to neoliberal governmentality, or in other words the ways in which presumably self-interested rational actors could be induced to alter behaviours towards conservation or other goals, in response to external, market or other forms of incentive structures. For PCCA, however, elements of truth governmentality also featured, as herders articulated their own priorities, and an understanding of 'right' human–nature relationships emerged. Practically, carbon sequestration, which has long underpinned the PV Standard, remained the central metric for issuance of credits and thus for flows of funds to the project. This was linked through CENTURY modelling to stocking rates and seasonal mobility between pastures. However, for local herders, carbon per se had little traction. Rather they were concerned to protect and better use pasture resources, to restore critical infrastructure, and find new ways to discharge what they felt to be their responsibilities for environmental protection and conservation.

The latter was realised in part through foregrounding diverse ontologies and values, as previously outlined and underpinned by affective socio-ecological relations. It is also notable that in this case non-monetary incentives – for example, participation in governance, coordination of labour and land rights, was at least as important to participants as additional financial incentives. Ongoing analysis of the PCCA scheme (Upton, forthcoming) is examining the specific impacts of the project on herders' livelihoods, pasture governance, carbon sequestration and on conservation. In governance terms specifically, the deployment of a PE lens facilitates critiques of carbon as a commodity, of the concept of ES and of neoliberal conservation. It does more than this however, in that it also enables examination of how neoliberal governmentalities operate, and may be shaped and contested through local agency. It thus enables us to go beyond blanket critiques of 'neoliberal natures' to understand intersections with other governmentalities, notably truth environmentalities and political ontologies, in specific places and contexts.

PEs of mining and resistance

Discoveries particularly of major copper and gold deposits have led to Mongolia recently being branded as 'a new mineral investment frontier'. By 2006, Mongolia's minerals sector accounted for 17% GDP and 58% of export earnings, rising to 30% and more than 80% by 2012 (World Bank, 2006). Subsequent downturns in global commodity prices have affected these GDP shares, but nonetheless a burgeoning mining sector is creating new governance challenges in Mongolia's rural areas. These pertain in part to loss of land rights and resource access, through both herders' physical exclusion from customary grazing areas, and through resource degradation, for example drying up of water sources. Mongolia's until recently extensive informal mining sector has also impacted on issues of resource, particularly water, quality and availability. A PE lens thus can illuminate how mining-induced environmental change intersects with the politics of resource access; in other words is pertinent to the first and the third of the questions characteristic of a PE approach, as highlighted above, namely 'how do nature and society influence each other and shape access to natural resources?' and 'what are the linkages between resource access/control, livelihoods and environmental change?' In the case of Mongolia, another dimension of PE is also evident; that of grassroots activism and the ways in which ostensibly political and social struggles may be underpinned by environmental change.

The herder-led Onggi River Movement (ORM), which emerged in 2001, is central herein. Its leader, Munkhbayer, who won the international environmental Goldman Prize (colloquially known as the 'Green Nobel') in 2007, was a local herder and chair of the *soum* (district) council. Initial mobilisations under his leadership used a variety of tactics to protest the drying up of local water courses and lakes, as well as pasture degradation and curtailment of access. The resistance grew rapidly and by 2006 ORM was one of 11 similar organisations in river basins across the country. Original activities and strategies reflected both practical concerns over land rights and access, but were also shaped by notions of 'traditional' values in relation to land and nature. Evident through examination of these mobilisations deploying a PE lens is the role of these political or diverse ontologies therein, and the central, underpinning importance of environmental change as a driver for activism. Contra to sovereign governmentalities, though which state actors sought to order and regulate mining practices and the behaviours of those affected by them, ORM and its partner organisations drew on truth governmentalities to contest resource rights and responsibilities. Specifically, ORM's early tactics included an 'ecology protest march', led by Buddhist lamas who officiated at a series of religious ceremonies at which prayers were offered for the 'preservation and protection of the Onggi ecosystem' (Upton, 2017). Other events organised by the MNPCM (Mongolian Nature Protection Civil Movement Coalition), which brought the various 'River Movements' together, also drew on Buddhist cosmologies as part of an assertion of particular, commonly held and local understandings of nature and the right relations and responsibilities of humans therein. These events, which may be deemed 'performative and ritualised moments of commonality' after Leach and Scoones (2007: 21) are highly symbolic and strategically deployed to underpin activism and mining resistance. They emphasise High's (2013: 754) argument that mining related politics are '... not necessarily delimited to the human realm', but rather non-human entities also feature. As argued elsewhere, evidence of the perceived agency of spiritual entities in this context may be seen in the recent growth of demands on Buddhist lamas to enact rituals to appease these spirits. Sources cited in High (2013) suggest that as many as 80% of Ulaanbaatar-based mining companies commission these kinds of rituals. Viewing mining related activism in Mongolia through a PE grounded political ontology lens is thus integral to understanding its emergence and particular manifestations.

These play out further in a number of ways and in relation to governance dilemmas. For example, ORM and its associated and successor organisations have long campaigned around particular aspects and absences of the state legislative framework. Colloquially known as 'the Law with the Long Name' (LLN), the

2009 'Law to Prohibit Mineral Exploration and Mining Operations in the Headwaters of Rivers, Protected Zones of Water Reservoirs and Forested Areas' has been central therein. Passed following pressure on the government by activists to protect water sources from adverse mining impacts, it was never implemented in full. It subsequently became pivotal to activists' political and governance strategies, with particular emphasis on the water resources essential for pastoralist livelihoods. Plans by the government to amend the Law, under pressure from mining companies, provoked a fresh round of protests in 2013, which landed some of the Movement leaders with lengthy jail sentences (Upton, 2017). What a PE lens helps to reveal here is intersections between sovereign governmentality through the LLN and truth governmentalities, including concepts not only of rights but of responsibilities vis a vis natural resources, whereby perceived failures by the government to act as 'guardians of environmental resources', were seen by members of activist movements as necessitating their intervention.

Thus, PEs of mining echo many of the issues highlighted by deployment of a PE lens across other governance sectors; namely in relation to carbon and conservation. These are summarised and wider lessons highlighted in the following conclusion.

Conclusions: rethinking PEs of governance

As a framework or loose body of theory PE has much to offer critical understanding of both empirical and conceptual governance dilemmas. Despite its lengthy history of highlighting struggles over resource access, use and values, and the power-laden relationships that underpin these, this potential has arguably yet to be fully exploited. In particular, post-structural PEs and emergent political ontologies, when taken together with more nuanced explorations of power in PE, for example in the work of Svarstad et al. (2018) and Fletcher (2017), have important contributions to make therein.

Through examination of three specific, and entangled, resource governance issues in Mongolia, centred on conservation, carbon and mining, this chapter has sought to show what a PE lens can bring to understanding and illumination of these issues. As argued above, a PE approach directs us to look at inclusion and exclusion in resource access and decision-making, through attention to issues of power and the ways these intersect with environmental change. In this way it offers some redress to other governance frameworks, which as highlighted in other chapters, may engage with power less directly. Recent PE work in this vein also seeks to address previous critiques from within the field, whereby arguably the operation of power in earlier PE was insufficiently theorised.

Through its post-structural turn PE also highlights the discursive processes that underpin these more legible issues of rights of access and decision-making. It enables examination of the ways struggles over representation and discourse reflect agency of diverse actors and how these then translate into inclusion or exclusion in both decision-making processes and resource access. In a related area, the recent interest in 'political ontologies' as a strand of contemporary PE, further enables critical understanding of intersections of power, diverse knowledges and their practical implications for governance. Non-human agency also begins to emerge in relation to governance and through this particular PE lens.

The deployment of a governmentality framework as part of a PE inspired exploration of power (after Fletcher, 2010), provides further explanatory insight and tools with which to explore governance dilemmas. As argued above, in the Mongolian case, this Foucauldian approach to power within PE illuminates the diverse and intersecting roles of discipline, sovereignty, neoliberalism and truth governmentality. Production of herders as 'conservationists', through for example donor or state programmes in Mongolia, could be seen as a form of 'green' disciplinary governmentality, wherein what constitutes desirable or conservation-oriented behaviour may be defined by external interventions, but is intended to be internalised and acted on by herders as subjects. Such disciplinary governmentality does, however, typically operate in conjunction with other forms, in this case sovereign governmentality, whereby natural resource governance and in this case particularly conservation, is shaped by legal frameworks and the exercise of state and donor power, with the two being mutually constitutive.

Many of the issues and concerns highlighted above are widely considered in relation to governance dilemmas and inform other frameworks; they are not solely related to a PE approach. The contention here, however, supported by examples from the Mongolian case study, is that a revitalised, multifaceted, contemporary PE is uniquely placed to enable critical understanding of the entanglements of power, agency, discourses and ontologies in producing and shaping governance practices.

References

Agrawal, A. (2005) 'Environmentality: community, intimate government, and the making of environmental subjects in Kumaon, India', *Current Anthropology,* vol. 46, no. 2, pp. 161–190.

Batbold, D., Batkhuyag, B., Ganzorig, B., Munkhnast, D., Munkhchuluun, B., Onon, Yo., Purevdorj, S., Sumiya, E., Selenge, G., Chimeddorj, B. and Enkhbayar, N. (2015) *'National Biodiversity Program 2015–2025'.* Ministry of Environment, Green Development and Tourism, Ulaanbaatar, Mongolia.

Blaikie, P. and Brookfield, H. (1987) *Land Degradation and Society.* Routledge, London.

Blaser, M. (2013) 'Notes towards a political ontology of 'environmental' conflicts', in L. Green (ed.) *Contested Ecologies: Dialogues in the South on Nature and Knowledge.* HSRC Press, Cape Town.

Brockington, D. (2002) *Fortress conservation: The preservation of the Mkomazi Game Reserve, Tanzania.* James Currey, Oxford.

Emerton, L., Erdenesaikhan, N., De Veen, B., Tsogoo, D., Janchivdorj, L., Suvd, P., Enkhtsetseg, B., Gandolgor, G., Dorisuren, H. C., Sainbayar, D. and Enkhbaatar, A. (2009) *The Economic Value of the Upper Tuul Ecosystem,* Mongolia Discussion Papers, East Asia and Pacific Sustainable Development Department. World Bank, Washington, DC.

Empson, R. (2011) *Harnessing Fortune: Personhood, Memory and Place in Mongolia*. Oxford University Press, Oxford.

Escobar, A. (2010) 'Postconstructivist political ecologies', in M. R. Redclift and G. Woodgate (eds) *The International Handbook of Environmental Sociology*. Edward Elgar Publishing, Cheltenham.

Fairhead, J. and Leach, M. (1996) *Misreading the African Landscape: Society and Ecology in a Forest-Savannah Mosaic*. Cambridge University Press, Cambridge.

Fernandez-Gimenez, M., Batkhishig, B., Batbuyan, B. and Ulambayar, T. (2015) 'Lessons from the Dzud: Community-based rangeland management increases the adaptive capacity of Mongolian herders to winter disasters', *World Development*, vol. 68, pp. 48–65.

Fletcher, R. (2010) 'Neoliberal environmentality: Towards a poststructuralist political ecology of the conservation debate', *Conservation and Society*, vol. 8, no. 3, pp. 171–181.

Fletcher, R. (2017) 'Environmentality unbound: Multiple governmentalities in environmental politics', *Geoforum*, vol. 85, pp. 311–315.

Fletcher, R. and Buscher, B. (2017) 'The PES conceit: Revisiting the relationship between Payments for Environmental Services and neoliberal conservation', *Ecological Economics*, vol. 132, pp. 224–231.

Goldman, M., Turner, M. and Daly, M. (2018) 'A critical political ecology of human dimensions of climate change: Epistemology, ontology and ethics', *WIREs Climate Change*, vol. 9, no. 4, e526.

High, M. (2013) 'Cosmologies of freedom and Buddhist self-transformation in the Mongolian gold rush', *Journal of the Royal Anthropological Institute*, vol. 19, no. 4, pp. 753–770.

Humphrey, C., Mongush, M. and Telengid, B. (1993) 'Attitudes to nature in Mongolia and Tuva: A preliminary report', *Nomadic Peoples*, vol. 33, pp. 51–61.

Kallis, G., Gomez-Baggethun, E. and Zografos, C. (2013) 'To value or not to value? That is not the question', *Ecological Economics*, vol. 94, pp. 97–105.

King, S. and Gurnell, J. (2005) 'Habitat use and spatial dynamics of Takhi introduced to Hustai National Park, Mongolia', *Biological Conservation*, vol. 142, no. 2, pp. 277–290.

Leach, M. and Scoones, I. (2007) 'Mobilising citizens: Social movements and the politics of knowledge', *IDS Working Paper 276*, IDS, Sussex.

Marin, A. (2010) 'Riders under storms: Contributions of nomadic herders' observations to analysing climate change in Mongolia', *Global Environmental Change*, vol. 20, no. 1, pp. 162–176.Ministry of Nature and Environment (MNE). (1996) *Biodiversity conservation action plan for Mongolia*, Ministry of Nature and Environment, Ulaanbaatar, Mongolia.

Ministry of Nature, Environment and Tourism of Mongolia (MNET), Global Environment Facility, Convention on Biological Diversity, United Nations Development Programme in Mongolia and Mongolian Nature and Environment Consortium. (2009) '*Fourth National Report on Conservation of Biodiversity in Mongolia (NRCBM)*'. Ulaanbaatar, Mongolia.

Nunan, F. (2015) *Understanding Poverty and the Environment: Analytical Frameworks and Approaches*. Routledge, London.

Peterson, G. (2000) 'Political ecology and ecological resilience: An integration of human and ecological dynamics', *Ecological Economics*, vol. 35, pp. 323–336.

Reading, R., Bedunah, D. and Amgalanbaatar, S. (2006) 'Conserving biodiversity on Mongolian rangelands: Implications for Protected Area development and pastoral uses', *USDA Forest Service Proceedings* RMRS-P39, pp. 1–17.

Reading, R., Murdoch, J., Amgalanbaatar, S., Buyandelger, S., Davie, H., Jorgensen, M., Kenny, D., Munkhzul, T., Onloragcha, G., Rhodes, L., Schneider, J., Selenge, T., Stotz, E. and Wingard, G. (2016) 'From 'Paper park' to model protected area:

The transformation of Ikh Nart Nature Reserve, Mongolia', *Parks,* vol. 22, no. 2, pp. 41–54.
Robbins, P. (2004) *Political ecology: A critical introduction.* Wiley-Blackwell, Chichester.
Robbins, P. (2012) *Political ecology: A critical introduction* (2nd ed.). Wiley-Blackwell, Chichester.
Robertson, M. (2012) 'Measurement and alienation: Making a world of ecosystem services', *Transactions of the Institute of British Geographers,* vol. 37, no. 3, pp. 386–401.
Schubert, J. (2005) *Political Ecology in Development Research. An Introductory Overview and Annotated Bibliography.* NCCR North-South, Bern.
Sneath, D. (2001) 'Notions of rights over land and the history of Mongolian pastoralism', *Inner Asia,* vol. 3, no. 1, pp. 41–59.
Sneath, D. (2003) 'Land use, the environment and development in post-socialist Mongolia', *Oxford Development Studies,* vol. 31, no. 4, pp. 441–459.
Sneath, D. (2014) 'Nationalising civilisational resources: Sacred mountains and cosmopolitical ritual in Mongolia', *Asian Ethnicity,* vol. 15, no. 4, pp. 458–472.
Stott, P. A. and Sullivan, S. (2000) *Political ecology: Science, myth and power.*Arnold, London.
Sullivan, S. (2017) 'What's ontology got to do with it? On nature and knowledge in a political ecology of the 'green economy'', *Journal of Political Ecology,* vol. 24, no. 1, pp. 217–242.
Svarstad, H., Benjaminsen, T. and Overa, R. (2018) 'Power in political ecology', *Journal of Political Ecology,* vol. 25, no. 1, pp. 350–425.
United Nations Development Programme (UNDP). (2007) *Sustainable Grassland Management Project Exit Strategy.* Unpublished report. Ulaanbaatar.
Upton, C. (2009) '"Custom" and contestation: Land reform in post-socialist Mongolia', *World Development,* vol. 37, no. 8, pp. 1400–1410.
Upton, C. (2010) 'Nomadism, identity and the politics of conservation', *Central Asian Survey,* vol. 29, no. 3, pp. 305–319.
Upton, C. (2012) 'Adaptive capacity and institutional evolution in contemporary pastoral societies', *Applied Geography,* vol. 33, pp. 135–141.
Upton, C. (2014) 'The new politics of pastoralism: Identity, justice and global activism', *Geoforum,* vol. 54, pp. 207–216.
Upton, C. (2017) 'Contesting development: Pastoralism, mining and environmental politics in Mongolia', in M. Watts, M. and L. Horowitz, L. (eds) *Grassroots environmental governance: Community engagements with industrial development.* Routledge, London.
Upton, C. (forthcoming) 'Conserving Natures? Co-producing PES in Mongolian Rangelands', *Development and Change,* Special Issue: 'Beyond Market Logics: Payments for Ecosystem Services as Alternative Development Practices in the Global South', in Shapiro-Garza, E., McElwee, P., Van Hecken, G. and Corbera, E (eds).
World Bank. (2006) *Mongolia: A Review of Environmental and Social Impacts in the Mining Sector.* World Bank, Ulaanbaatar.
Zhang, M., Borjigin, E. and Zhang, H. (2007) 'Mongolian nomadic culture and ecological culture: On the ecological reconstruction in the agro-pastoral mosaic zone in Northern China', *Ecological Economics,* vol. 62, no. 1, pp. 19–26.

9

THE STATE AS A PERSON

The role of interface bureaucrats in everyday natural resource governance

Mikkel Funder

Introduction

How should we understand the agency of state actors in natural resource governance? The state and its policies and laws clearly play a key role in natural resource governance. At the same time, however, it has become increasingly clear that the state is no monolith: state actors are varied and operate in widely different contexts, their reach is often constrained, and their authority is challenged and mimicked by a range of other institutional actors. State actors are, furthermore, not only institutional actors but also individuals with personal lives, perceptions and interests. What does this mean for our understanding of state actors and their agency in natural resource governance, and what does it imply for the way national and international policies and plans play out in practice?

This chapter discusses the roles, relationships and practices of a particular group of state actors which have, until recently, gained only limited attention in the literature on the governance of natural resources and the environment in the Global South – namely, the local managers, technical specialists and extension agents who implement state policies, laws and projects on the ground. Often assumed to be mere extensions of the central state, a growing body of literature is documenting how these 'interface bureaucrats' – also known as street level bureaucrats – play a critical role in shaping natural resource governance in practice.

The main aim of the chapter is to encourage a grounded view and analysis of local state actors and their practices in everyday environmental governance. Drawing on selected literature on the state and environmental governance, the chapter discusses how we should understand the role of interface bureaucrats in natural resource related interventions. Specifically, it discusses five factors that are key to understanding the rationales and practices of interface bureaucrats, namely, their role as frontline representatives of the state; the local institutional and political dynamics; their subjectivities; their professional and social networks; and their informal practices. In so doing, the chapter especially draws on the developing literature on the anthropologies of the African state (Bierschenk and Olivier de Sardan, 2019; Blundo and Le Meur, 2009; De Herdt and Olivier de Sardan, 2015) but seeks to link this to other work in public administration studies, organisation studies, development studies and anthropology.

To illustrate the discussion empirically, the chapter draws on the author's involvement in research on the everyday implementation of climate policies and projects in Zambia. This work showed that climate change adaptation policies and projects are vested with political interests, including efforts by central state actors to legitimately expand control over people, land and modes of production (Funder et al., 2018). However, it also showed that when interface bureaucrats go about implementing these polices and projects in practice, they change them significantly (Funder and Mwemba, 2019). This chapter discusses the factors that influence the rationales and actions of interface bureaucrats in such situations.

Zooming in on interface bureaucrats – and why they matter

In his well-known discussion of how a state 'sees' the world, Scott discusses how state interventions can be understood as efforts to make people and places administratively manageable and subject to the expert solutions of High Modernism (Scott, 1998). Li explores similar dynamics, but points to the co-produced nature of the 'improvement schemes' through which governmentality is exercised (Foucault, 2007; Li, 2005, 2007b). She shows how state actors, development agencies, non-governmental organisations (NGOs) and academics may come together in the construction of environmental discourses such as 'Community Forestry' which seek to align otherwise conflicting interests (Li, 2007a; Sareen, 2016). State policies and programmes in natural resource management and environmental conservation – whether inclusive and community-based or not – are thus deeply entangled in the efforts of the state and other actors to govern territory, resources and behaviour (Bluwstein and Lund, 2018; Death, 2016; Lindegaard, 2018).

Yet state interventions do not take place in a vacuum. Actors respond to and interact with them. Scott's own work famously emphasises how peasants often evade or undermine interventions through everyday resistance such as foot-dragging and false compliance, rather than overt rebellion (Scott, 1985). Other bodies of work in and around legal anthropology have shown how farmers and other local resource users creatively engage and negotiate the formal and informal rules and procedures that govern access to natural resources (Benda-Beckmann, 1981; Berry, 1993). At the same time, several studies have highlighted how a variety of actors outside the conventional notion of 'the state' – for example, local community groups, NGOs and chiefs – make competing claims to public authority or carry out functions and practices that imbue them with 'stateness' (Lund, 2006; Lund and Boone, 2013). Local actors thus not only resist state interventions, but may also actively engage and reshape them, or even seek to 'become' the state.

Critical studies of natural resource governance in development have tended to focus either on the interests of global and national actors who devise and impose policies and interventions – or on the communities who are subjected to such interventions. Less interest has been afforded to the everyday practices, positions and subjectivities of local state practitioners, that is, the government staff who implement the policies and interventions on a daily basis.

In the broader academic literature, a number of different terms have been applied to describe these actors, of which Lipsky's 'street level bureaucrats' is perhaps best known (Lipsky, 2010). Here, I use the term 'interface bureaucrats' as applied in the recent work on the anthropologies of the African state, which – through its emphasis on interface – emphasises the relational aspect of their position (Olivier de Sardan, 2015). In the field of natural resource governance, such interface bureaucrats are typically employed as the local level staff of sectoral ministries and associated line agencies within agriculture, water, forestry, environmental management, community development, disaster management and so on. In many African countries, their agencies are either deconcentrated branches of the central state or have a more devolved function in which they support or even work for local governments. Their positions range from field extension agents, via technical subject matter specialists, to the senior staff who manage and coordinate their agency's work at district level or similar.

Interface bureaucrats are of analytical interest for several reasons. First, they are situated at the heart of the state–citizen interface and thereby provide a window into the practical, everyday embodiment of state actors and their interactions with other actors in natural resource governance. Second, through their role as everyday regulators and implementers of laws, policies and programmes, local state actors have an important – though often forgotten – influence on the way natural resource interventions are carried out in practice. This makes them important actors to understand better, especially in light of the growing interest in the 'everyday governance' of natural resources and the environment (Cornea et al., 2017).

Lastly, the everyday practices and agency of local state actors can provide important insights into the ways in which institutions for natural resource governance are shaped and change. Developing work in Critical Institutionalism has examined the ways in which actors piece together institutions and arrangements for natural resource governance through processes of bricolage (Cleaver, 2012; Cleaver and de Koning, 2015). This highlights the importance of understanding the everyday micro-processes through which actors shape and affect institutions and interventions. Much of the work on bricolage has – for good reasons – focussed on communities, but studies of local state actors can help deepen our understanding of the interplay of different actors in this respect. It can, furthermore, provide opportunities for fruitful engagement with related work in other fields, such as the studies of 'institutional work' and embedded agency in organisation studies (Lawrence et al., 2013; Zietsma and Lawrence, 2010).

Significantly, an interest in the practices and everyday agency of local state actors does not imply a naïve approach, or a leaning towards methodological individualism. Rather, it is an attempt to understand how interface bureaucrats are situated in a local web of relations and interests, how this ties in with their practices and interactions with other actors, and what the outcomes are.

In the following I discuss five factors that are key to understanding the work of interface bureaucrats in natural resource governance, namely, (i) their role as frontline representatives of the state, (ii) the local institutional and political dynamics, (iii) their subjectivities, (iv) their professional and social networks, and (v) their informal practices. Of these, some have been well explored empirically, while others have received little attention so far. The discussion is illustrated with findings from research on the climate change interventions in Zambia and seeks to draw on other studies from the developing literature on the topic.

The role as frontline representatives of the state

How should we understand the tasks of interface bureaucrats as frontline representatives of the state? Any in-depth look soon shows the complexity of their roles. This is partly reflected in their official mandates. A district forest officer may find herself arresting illegal charcoal burners in the morning and facilitating community-based forest governance in the afternoon. She is responsible for administering the state's monopoly on violence, but also responsible for the delivery of public goods and services (Blundo and Glasman, 2013).

In addition to this, the role and position of interface bureaucrats also has more implicit purposes. As frontline representatives of the state, they are engaged in its ongoing (re-)production. This includes the work of making people and territories 'governable', that is, bringing them under state control and making them subject to administration (Scott, 1998; Vandergeest and Peluso, 1995). In extension of this, interface bureaucrats play an important role for the central state in pacifying and containing local grievances, thus avoiding the involvement of central government and the risk that local conflicts become national. These roles are not new in African settings, but reach back to colonial times (Berman, 1992).

Interface bureaucrats are, furthermore, important actors in the ongoing discursive construction of the state and the state–citizen relationship (Gupta, 2006). Citizens do not passively accept state interventions but seek to resist, engage and reshape them. Moreover, in many African settings, other institutional actors contest the legitimacy of the state to intervene or compete with it for public authority. Interface bureaucrats thus play a key role in extending and producing the legitimising strategies (Sikor and Lund, 2009) of the central state, whereby it seeks to legitimise its authority to govern natural resources and intervene in their use and management.

The official and unofficial roles of interface bureaucrats are thus closely linked to reproducing state authority, both explicitly as regulator and service provider, and implicitly as the symbol of public authority and territorial control. In this light it is tempting to assume that the interests and actions of interface bureaucrats can be understood – perhaps even predicted – through a scrutiny of central state policies and interests alone. However, this would overlook the real-life complexity of the work life of interface bureaucrats.

First, interface bureaucrats operate with a high degree of discretion vis-à-vis official laws, rules and procedures. Lipsky's US-based work showed how the position of public servants at the frontline of government provided them with considerable discretionary space in the implementation of policies and interventions, and that they actively employed this to devise their own interpretations and rules, thereby effectively becoming policy makers (Lipsky, 2010; Weatherley and Lipsky, 1977). Other recent work in western settings has arrived at the same conclusions, including studies of environmental interventions (Crewett, 2015; Lind-Riehl et al., 2016; Maier and Winkel, 2017; Ricks, 2016; Sevä and Jagers, 2013). Coming from a different direction but exploring similar themes, research on the anthropology of the African state has highlighted the 'practical norms' of interface bureaucrats, which deviate from the official norms of the bureaucracy (Bierschenk and Olivier de Sardan, 2014; De Herdt and Olivier de Sardan, 2015).

Second, the implementation of laws and policies is a social process that requires interface bureaucrats to relate to and interact with other actors. In Public Administration studies, implementation theory has long since challenged the notion that interventions are merely an effect of the politics of policy making (Barrett and Hill, 1984). This work thus emphasised that while power is crucial in public interventions, it also necessarily involves bargaining, compromise and discretion, to the extent that implementation outcomes can be seen as 'negotiated order' (Barrett, 2004). In development studies, the work of Long and others showed how the interaction between extension agents and community members could be seen as an encounter of different lifeworlds and interests, which were nevertheless partially dependent on each other and thus co-produced outcomes (Mosse, 2005; Long, 1989).

These bodies of literature add important dimensions to our understanding of local public officials in natural resource governance. They point out that interface bureaucrats are not mere 'implementation robots' but have a degree of semi-autonomy when carrying out their work; and they indicate that the task of making people and places governable is by no means straightforward and may involve compromise and negotiation.

Local institutional and political dynamics

When carrying out their work, interface bureaucrats rarely have the physical reach or hegemonic power that is sometimes assumed. Most interface bureaucracies operate with severe resource constraints in terms of insufficient staff and operational funds, which greatly limits and complicates the practical scope of their operations (De Herdt and Titeca, 2016; Hamani, 2014). In our Zambian study areas, district agricultural agencies typically received only some 25% of the funds requested in work plans, and even these are often substantially delayed or never turn up in full. Fuel and spares for motorbikes and other vehicles used by extension staff are in perennial short supply, significantly limiting the extent and frequency of their physical reach in the typically vast areas that they are required to cover. Clearly, the power to enforce regulations through use of armed force – as in the case of, for example, wildlife rangers in protected areas – is a significant instrument of domination. However, even in such cases the everyday physical reach of interface bureaucracies is limited, as witnessed by the frequent inability of wildlife agencies to prevent everyday poaching and transgression of protected area boundaries.

In the African context, interface bureaucrats, furthermore, operate in settings where the legitimacy of the state is often contested or circumvented by competing institutional actors. This includes customary authorities who claim authority as managers and allocators of land and water rights, and as 'developers' in their own right – regardless of whether they are colonial constructs or not (Ranger and Hobsbawm, 1983). In many African countries local governments are, furthermore, seeking to establish greater authority vis-à-vis the local line agencies of the central state – calling for greater devolution and claiming more control over natural resources and associated revenues (Ribot, 2007). Meanwhile, NGOs and development agencies implement projects and provide services that potentially threaten the legitimacy of interface bureaucrats as knowledge experts, and may lead to the emergence of 'project law' in competition with statutory laws (Weilenmann, 2017). In a similar vein, community organisations in some settings take on local state-like functions, such as policing or natural resource monitoring (Funder et al., 2013; Telle, 2014). In their relations to interface bureaucrats, these different actors variously distance themselves from the state or compete with it for public authority (Lund, 2006). This in turn opens a degree of space for individual community members to shift between different institutional actors when seeking support for their claims and interests, thereby, in turn, also adding or detracting from the legitimacy of the actors in question.

The relative power and authority of interface bureaucrats in these complex institutional landscapes will, of course, vary between and within countries, and over time. In Zambia, state agencies had relatively strong control in many rural areas in the decades following independence, but with the fall of the one-party state a more fragmented picture emerged, including the re-emergence of chiefs as a strong political force (Haller, 2012). Subsequent structural adjustment programmes imposed by the IMF meant that civil services were greatly cut back, with significant consequences for the staffing allocations of extension-oriented agencies in, for example, forestry, water services and agriculture. In Kenya, the ongoing devolution process has meant that formerly weak local governments now have a degree of actual planning and decision-making powers and not insignificant budgets (Cheeseman et al., 2016). Interface bureaucrats in some state agencies – such as agriculture, water and health – are therefore having to reorient their roles and relationships with local governments.

Rather than a static relationship, the position of interface bureaucrats vis-à -vis other local institutional actors and communities can thus be seen as a dynamic and shifting 'force field' (Nuijten, 2005). This does not by any means suggest that interface bureaucrats are somehow powerless. What it does mean is that their role and authority as representatives of the state – as an organisation and as an idea – needs continuous work to be maintained, and that this work requires considerable pragmatism and discretion in order to succeed within the settings where they operate.

The overlooked role of subjectivities

So far, much of the work on interface bureaucrats in African settings has centred on practices of corruption. But while this is certainly prevalent, an exaggerated emphasis on rent seeking risks drowning out other features of their professional agency and identity. The subjectivities and lifeworlds of interface bureaucrats are particularly under-researched, although some recent work has begun to examine the topic (Hahonou, 2015; Le Marcis and Grard, 2015).

How do interface bureaucrats themselves perceive their roles? In Zambia, interface bureaucrats described their role as one of helping communities transform into better, more developed places. Community members were perceived as unfortunate victims of a harsh environment and outdated production systems, or as one staff member said it, 'good people in a bad place'. A common narrative was that community members were trapped in vicious circles of poverty and environmental degradation, compounded by climate change and other factors. Interface bureaucrats saw themselves as central actors in addressing these challenges through the provision of technical expertise and enforcement of sound law and policy.

Such notions are well in line with the discursive 'rendering technical' of local livelihoods and their environment as discussed by Li, which helps to legitimise the need for external expertise and associated interventions (Li, 2007a). Yet these discourses are not simply 'out there' but are engrained in their subjectivities. One aspect of this is the organisational sense-making through which they seek to make meaning of their professional role and tasks (Weick, 1995). This sense-making is partly founded in the official visions and training materials of their parent ministries, which emphasise the developmental role of their staff, and tallies well with the 'state vision' discussed by Scott (1998). However, such an analysis only shows part of the picture.

Firstly, during our interviews, interface bureaucrats expressed sense-making rationales particular to their position at the frontline of the state. District staff in the agriculture, forestry, wildlife and community development agencies saw their local and experience-based knowledge as a key asset and defining feature, which set them apart from their colleagues in their parent ministries in Lusaka – who were often described as having no understanding of the issues on the ground. The fact that staff at ministerial level often have a past as interface bureaucrats themselves was explained as a result of the way people change when they take up jobs at higher levels, losing sight of what matters in practice. Interface bureaucrats thus saw themselves as critical actors in aligning misguided and/or generic national policies with local realities, by bringing their understanding of the local sociopolitical, institutional and biophysical context into play. Hence while they considered themselves very much as representatives of the state, it was as a distinct group of state actors, whose actions were important for the state to function in the first place. This does not mean that they have no ambition of advancing in the hierarchy themselves but should be seen as a means to make sense of their current position and role in the state and in life. These perceptions align well with those expressed by interface bureaucrats elsewhere, including in the western context (Hupe and Hill, 2007).

These rationales have implications for the actions of interface bureaucrats. In Zambia, government staff working in agriculture, water, forestry and community development frequently employed their discretionary space to reinterpret, adjust and sometimes significantly change the nature of policies and projects. I return to this, but the point here is that interface bureaucrats expressed a strong belief that this work was quite legitimate for them to do. While it could not be stated officially in reports or conversations with their superiors at other levels, it was an inherent part of their role as those who knew the local context and dealt with the actual issues of development and resource management. Rather than attempts to undermine the state as such, their discretionary reorientation of national interventions can thus be seen as palliative efforts to make the state work (Olivier de Sardan, 2014). Significantly, however, this also asserts their professional identities and associated sense-making. The very act of adjusting or changing national policy thus creates the distinction between them and other parts of the state. By way of example, interface bureaucrats in Zambia sometimes joked with colleagues or NGO staff that their superiors in Lusaka had no idea what they were doing.

Second, the individual lifeworlds of interface bureaucrats extend well beyond the state and influence their rationales. The sight of a lone government officer stuck behind his desk in a simply furnished rural district office might lead to assumptions that his lifeworld is bounded by the particular locality in which he works and defined only by his role as a state representative. Yet interface bureaucrats – particularly in the higher ranks – are often well travelled and have resided in several locations around the country during their postings and training. They have diplomas and degrees, follow world news on radio and TV, and communicate with friends and family in the capital and abroad via active social media accounts. They, furthermore, harbour political convictions, religious beliefs and personal dispositions.

This has bearings on their work-related actions and practices. In Zambia, a community development officer recounted how her strong faith compelled her to visit very poor households more often than better off ones, although this was against regulations and the subject of occasional complaints from other community members. An agricultural extension officer described how he had been indifferent to climate change until watching 'An Inconvenient Truth' on satellite TV. He then became captivated by the subject and – in the absence of appropriate government guidelines at the time – used Internet searches to read up on climate resilient farming techniques for use when providing extension advice. Although such accounts clearly have an element of self-representation, interviews with community members and co-workers appeared to confirm the actions in question.

Third, and in extension of this, many interface bureaucrats must deal with the personal frustrations of working in an exposed position where results are often difficult to deliver. As Stepputat and Nuitjen (2018) have pointed out, surprisingly little attention has been paid to the role of emotion and feeling in theories and studies of the state, despite its potential significance. This is also relevant in the study of interface bureaucrats. The chronic shortage of resources and often contested nature of their work, and the complex social and institutional relations in which they are embedded, means that their work life is rarely straightforward, and often experienced as frustrating.

The everyday practices of interface bureaucrats thus also have an emotional dimension, and associated personal coping strategies (Diarra and Ousseini, 2015). During interviews, an agricultural extension worker expressed frustration at the persistently late delivery of seeds and fertiliser packages as part of a national agricultural programme. Although beyond his control, community members criticised him for this. As a result, he would avoid going to communities during such times. Along similar lines, studies of health workers in West Africa show how they experience a disenchantment with the public health system of which they are part, and that this in itself becomes a rationale for engaging in petty corruption (Diarra and Ousseini, 2015; Hahonou, 2015).

Professional and social networks

Like all organisational actors, the everyday work of interface bureaucrats takes place in a web of professional relationships and broader social networks that both enable and constrain their work. Hupe and Hill thus suggest a need to understand better the 'micro networks' of local public servants (Hupe and Hill, 2007). In the African context, Blundo discusses the multiple 'accountabilities' of interface bureaucrats, that is, the formal and informal professional and social relations in which they are embedded (Blundo, 2015). In the following I sketch out three aspects of this, namely, the relationships between interface bureaucrats and (i) their superiors, (ii) their peers and (iii) the broader social networks of which they form part.

Relations to superiors at higher administrative levels

An important part of the work life of interface bureaucrats evolves around managing interpersonal relations within the state apparatus itself. This includes relations to senior staff in higher tiers of government at provincial and national levels. Maintaining such relationships has both informal and formal dimensions. Patronage and nepotism can play an important role in promotion, and when available such upward relations are important for interface bureaucrats to maintain. Yet as Therkildsen has shown, simplistic perceptions of recruitment and career advancement in African bureaucracies must be avoided (Therkildsen, 2014). Official procedures such as regular progress reporting on annual work plans do matter to superiors at provincial and national levels, providing as they do a routinised practice through which hierarchy is continuously reproduced.

This is also the case in Zambia. Though often lamented, upward reporting and similar routines that involve engagement with higher levels of the administration is an important means for interface bureaucrats to maintain a relationship to superiors and maintain a clean track record. However, interface bureaucrats in the natural resource sectors typically have a high degree of freedom in *what* they actually report to their superiors – as long as action is evident and in line with overall policy. While the use of funds is subject to some degree of scrutiny, the reported results and outcomes of service delivery and everyday law enforcement is rarely validated in practice by higher administrative levels. One Zambian agricultural officer, speaking of his superiors at Provincial level, put it thus: 'They know it's impossible [...to check everything...] they just want to see that something is going on'.

Moreover, superiors at national and provincial levels often do not *want* to be involved in local routines and problems. Apart from taking up their time, it potentially entangles them in local conflicts and implementation failures. Hence while interface bureaucrats are regularly in contact with their superiors on practical matters, it is often in the interests of both parties that the latter are left out of the equation when it comes to the complexities of everyday policy implementation.

As Kairu et al. (2018) have shown in the Kenyan context, one effect of this relationship is that progress reporting to higher levels typically becomes a performance – that is, an important but staged process in which interface bureaucrats make an effort of reporting to higher levels that which is expected, while in everyday practice doing things quite differently. Below, I discuss a case of this from the Zambian context. A further effect is that interface bureaucrats carefully regulate the degree to which superiors and other actors at higher administrative levels are called on to deal with problems, conflicts or sanctions. Not involving superiors represents a risk if the matter develops out of hand, yet in many cases this is the preferred option. Apart from the unwillingness of superiors to be involved in the first place, bringing them into the picture may lead to exposure of mistakes, or require that unwieldy official procedures and rules are brought into play on matters that were better solved through informal measures.

Relations to peers

Peer relationships in African interface bureaucracies have received little attention yet form an important part of their ability to function. The staff who work together within a particular agency at, for example, district level are often no more than a handful, and careful management of personal and professional relations to one's colleagues is required (Blundo, 2015). This is partly necessary for practical reasons; the scarce resources and frequent field-based activities mean that mutual favours and ad hoc coordination is continuously necessary – such as sharing the only available vehicle, or asking a colleague to set up a meeting in a community that he happens to pass by.

Peer relations play an important role in the reproduction of the informal practices of interface bureaucrats. In Zambia, interface bureaucrats described how they as new staff had learned the informal practices by experiencing how their colleagues addressed a given situation. It is, furthermore, through the (often silent) acceptance of one's colleagues that discretionary practices are sustained. Sometimes, this takes on overt forms. In Kenya, environmental officers made sure to introduce new staff to informal, unwritten agreements that had been struck with local communities on natural resource governance (Funder and Marani, 2015).

Peer relations across different agencies and ministries add a further dimension to these dynamics. A variety of different agencies typically work within agriculture and natural resource management, and although their target groups and physical work space often overlap, they have different histories and asymmetric power relations. Some have long trajectories as key agencies in state resource control and revenue generation, such as forestry services and mining. Others are more recent constructions with a weaker power base, such as community development agencies or the environmental management agencies developed in recent decades. Some – such as Disaster Management agencies – have the authority to convene other agencies, while others do not.

This has a bearing on the room for manoeuvre of interface bureaucrats. Staff in powerful agencies can dominate local planning procedures and call on stronger political backing when necessary. Staff in weaker agencies must navigate more carefully and seek alliances with stronger agencies or develop alternative means of carrying out their work. Funder and Marani (2015) described how a Kenyan environmental officer arrived at the district headquarters as the first representative of a newly formed environmental agency. Seen as a potential threat by staff in some other local agencies, and with very little organisational clout and resources at his disposal, he developed an informal network of sympathetic colleagues, community-based organisations (CBO) representatives and community members. Through this network, he was able to begin the (highly pragmatic) implementation of the environmental law.

Position in local social networks

The professional relationships of interface bureaucrats form part of the broader social networks within which interface bureaucrats are embedded. Olivier de Sardan has discussed the 'solidarity networks' of government staff in African countries, that is, the multiple networks of belonging within which interface bureaucrats are positioned (Olivier de Sardan, 1999). These may include family members and friends but also former class mates, university peers, neighbours, or members of the same church or party. Members of such networks may have expectations that interface bureaucrats will service their needs (provide special favours, reduced fines, etc). Such expectations complicate the work life of interface bureaucrats and may place them in precarious situations. In Zambia, a member of the district administration recounted how he had been inundated with requests for food relief from his home village during an extreme flood, although the village wasn't actually located in the district where he worked. When he responded that he was unable to help, he was accused of not looking after his own.

At the same time, such relationships can obviously also provide opportunities for interface bureaucrats, including the benefits of petty corruption. Importantly, informal social networks also provide an important means for interface bureaucrats to carry out their work tasks. A common but problematic notion of the staff in deconcentrated branches of the state is that they are somehow foreign and disconnected from citizens in the communities where they work. In practice, however, their social networks often extend well into communities (Anders, 2009). In this respect, interface bureaucrats working in areas where they were born and raised can be at a particular advantage; discussing the 'vicinal accountabilities' of local forest staff in Niger, Blundo describes how staff originating in their place of work appear to experience fewer offences by community members than staff members from elsewhere (Blundo, 2015).

Regardless of their own origin, interface bureaucrats often rely heavily on reciprocal relationships with community members in carrying out their work (Long, 1989). This includes community 'gatekeepers' (e.g. village heads), as well as broader groups of individuals who can pave the way for implementation and compliance and add legitimacy to the role and actions of interface bureaucrats. Likewise, community members seek to benefit from informal relations with interface bureaucrats. As Nuitjen has pointed out, academic interest in how community members resist and reshape state interventions should not blind us to the fact that in many cases, community members actively *seek* the benefits provided by the state (Nuijten, 2005; Stepputat and Nuijten, 2018). Establishing informal working relationships with interface bureaucrats such as agricultural extension officers, water management staff etc. can therefore be important for community members.

In such relationships, the power is obviously skewed towards interface bureaucrats – but they also entail necessary compromises. Our research in Zambia showed how interface bureaucrats sometimes accepted that community members steered climate change adaptation activities in directions that were different or even contradictory to government policy. Even if interface bureaucrats did not like this, they saw it as a necessary trade-off in order to avoid community members resistance to interventions or turning to competing authorities such as Chiefs (Funder and Mwemba, 2019).

Informal practices

In the above discussion, I have provided brief examples of the informal practices employed by interface bureaucrats in their everyday roles as frontline representatives of the state. In the following I look more closely at this and provide an extended empirical illustration.

A central concept in the recent anthropological literature on African state is that of practical norms, understood as 'the latent regulations of the practices of civil servants when these do not follow official regulations' (Olivier de Sardan, 2015: 21). Practical norms do not replace the official juridical and ethical norms of a bureaucracy but exist alongside them. They are sometimes made explicit but are more often tacit and unspoken.

Intended mainly as a heuristic device, the concept of practical norms is not extensively developed in theoretical terms. However, it can help us avoid the pitfalls of methodological individualism, by highlighting that the informal and discretionary practices of interface bureaucrats are more than just random independent actions. Linking to Bourdieu, one might say that the practical norms are engrained in the interface bureaucrat's habitus and articulated in an intuitive feel for the game (Bourdieu, 1990).

Olivier de Sardan (2014, 2015) discusses different types of practical norms. This includes 'adaptative norms' (minor tweaks and bending of rules); 'semi-tolerated norms' (generally accepted diversion from official norms); 'transgressive norms' (significantly beyond or opposed to official norms, e.g. corruption) and palliative norms.[1] The latter are also well beyond official norms, yet are focussed on maintaining their 'spirit', that is, they seek to make the state and its functions work in a context where it is not possible via other means.

While this typology is obviously schematic, it is helpful in pointing out that some practical norms are little more than oil to make the machinery function, while others diverge more fundamentally from the official norms. It is the latter, I would suggest, that are particularly interesting. How do these translate into practices in everyday environmental governance? In the following I focus particularly on palliative practices.

In Zambia, interface bureaucrats struggled with three key dilemmas when seeking to implement government policies and interventions on climate change adaptation. These were as follows:

i Intervening with limited reach, that is, how to intervene with very limited staff and budget in a context of fragmented institutional authority.
ii Implementing generic policies, that is, how to implement policies that were based on universal assumptions that did not match local social, institutional and ecological realities.
iii Managing conflicting interests, that is, dealing with the multiple and often contradictory interests of other actors as they seek to counter or engage state interventions.

While the particular manifestation of such dilemmas depends on the setting in question, they echo the dilemmas described for interface bureaucrats elsewhere (De Herdt and Titeca, 2016; Hamani, 2014; Lipsky, 2010). Intriguingly, they also resonate with Berman's discussion of colonial district officers in Kenya, and their contradictory role of being both an agent of external domination and an agent of local accommodation (Berman, 1992).

In order to address these challenges, interface bureaucrats in Zambia applied the following overall practices:

a Seeking to intervene with limited reach, they informally delegated responsibility and authority to selected community members. This exceeded and crafted informal agreements and arrangements with them.
b In order to implement generic policies, they 'staged' project activities that allowed them to report implementation progress to their superiors at provincial and national level – while discretely prioritising other climate change adaptation activities that they consider more appropriate.
c In order to manage conflicting interests, they negotiated solutions with communities and other institutional actors, rather than asserting their official authority. They followed the path of least resistance and implemented the policy in different ways and to a different extent according to what was possible within the power dynamics of a given setting.

Again, such 'inventive practices' (Hamani, 2014) play out differently depending on the given setting, but their overall thrust and the pragmatism they entail is similar to practices elsewhere, both in African settings and beyond (Blundo, 2014; Kairu et al., 2018; Sevä and Jagers, 2013). In the following, I provide a more detailed example of how one interface bureaucrat in Zambia applied such practices, and how his subjectivity and networks came into play.

Side-lining conservation agriculture – a case

Conservation agriculture is a set of farming techniques that aims to preserve water and nutrients through minimum soil disturbance, permanent soil cover and diversification of plant species. Although the approach has long been an element of agricultural extension activities in Zambia and elsewhere, it has recently been promoted by a variety of UN organisations, donor agencies and NGOs as a so-called 'climate smart' strategy whereby farmers can adapt to reduced rainfall and increasing temperatures (Arakelyan et al., 2017; Whitfield, 2016). As a technological intervention in household farming practices, conservation agriculture can be seen as an example of how climate change adaptation measures can be employed to legitimise and depoliticise external interventions in rural production (Taylor, 2018). However, as our research in Zambia showed, there is a further dimension to this.

Conservation agriculture is central component in Zambia's climate change adaptation strategy, which district agricultural offices are expected to show progress on. However, many farmers dislike the approach, claiming that the approach reduces yields. Moreover, in our study areas many agricultural staff members are themselves highly sceptical. Based on their personal experience and past efforts to implement the approach, they find that the outcomes of the techniques employed are highly dependent on a number of context-specific ecological factors.

In one of our study areas, the leading agricultural officer had similar concerns. Having worked in the area for many years, he believed that the sandy soils of southern and western Zambia were unsuited for conservation agriculture and that the practice actually reduced crop yields compared to conventional approaches. During interviews, the agricultural officer lamented the fact that he was tasked to implement a policy that neither he nor local farmers believed in. He explained that it threatened his authority as a knowledge expert among local farmers and referred to a mounting critique by farmers and local councillors against the emphasis on conservation agriculture in agricultural extension work. He also felt that it undermined his long-term efforts to introduce efficient, high output farming systems through crop diversification, improved seed varieties and improved water management – approaches that he believed strongly in.

Initially, he sought to convey this experience through the official planning process. When submitting workplans and budgets to the provincial agricultural office, he argued that conservation agriculture was inappropriate for this particular area, and instead proposed a greater emphasis and budget allocation for other means of climate change adaptation – such as early maturing maize varieties and small-scale water harvesting infrastructure. In all cases these proposals were turned down by the provincial level, with the explanation that conservation agriculture was national policy and should be followed. During interviews, the agricultural officer expressed frustration at this yet also explained that it was no surprise and that it was a '…condition of work in this system'. He said that blanket policies from the central government was a continuous challenge, and that his ministry was lucky to have people like him and his colleagues to apply 'real knowledge'.

Having failed to secure support from the provincial level, the agricultural officer thus resorted to other means. Drawing on his own social networks and those of his extension officers, he liaised with selected influential community members – including CBO representatives and Headmen – to establish a number of conservation agriculture 'demonstration plots' in selected communities. The plots were cultivated using conservation agriculture methods by voluntary 'lead farmers'. While such plots are a well-established method of knowledge diffusion in agricultural extension work, they were in this context little more than staged activities: they provided tangible showcases that could be reported on to the provincial and central level, thereby satisfying reporting requirements and showing that, although uptake of conservation agriculture among farmers was low, something was being done to implement national policy.

The demonstration plots were thus essentially a performance of policy implementation (Kairu et al., 2018). Two things are, however, important to note.

Firstly, there was more to this practice than mere representation (Mosse, 2005). The leading agricultural officer thus went beyond staging events, and actively sought to promote *other* alternative climate change adaptation practices that he did believe in. He informally instructed extension workers to focus on promoting alternative farming practices such as new crops and seed varieties and sought to develop small-scale irrigation schemes. By using his discretionary space, he allocated staff time to support such practices, and facilitated NGO funding to them. In so doing, he was not seeking to undermine the authority of the state as such. During interviews, he emphasised his strong belief in the state as developer and service provider but explained that – given the blanket policies imposed by 'administrators and politicians' at central level – it was necessary to reorient the direction of national policies to comply with the reality of the ground and the location-specific knowledge and expertise of himself and his colleagues. Other interface bureaucrats in the area expressed similar perceptions and engaged in similar practices, thus reflecting – in Olivier de Sardan's terminology – palliative rather than transgressive norms. It is notable, however, that the efforts of the agricultural officer and his colleagues appear to include a distinct attempt to 'improve' on the state and its policies rather than simply reproducing it. This indicates a need to understand interface bureaucrats as more than mere 'fixers' of dysfunctional policies and organisations.

Second, the conservation agriculture demonstration plots were essentially co-produced. The lead farmers involved were well aware that the plots were not taken seriously by the agricultural staff. Their main production came from other, conventionally cultivated plots. Their interest was not so much in the CA practices themselves, but in the benefits that came from being lead farmers as such. This included access to seedlings for alternative crops, and a role as state-appointed knowledge experts which provided social leverage in the community. One lead farmer even used his demonstration plot to document that CA produced smaller yields than conventional practices and shared the results with other farmers and extension officers – thereby boosting his status as a resident expert in farming techniques. The basis for the practices carried out by the interface bureaucrats in our study areas were thus to a large extent dependent on active engagement and coproduction by community members.

Conclusion

The intention of this chapter has not been to portray interface bureaucrats as unsung heroes, or to discount the role and power of international actors and the central state in environmental governance. The latter actors unquestionably play a central role in shaping not only the specific policies that interface bureaucrats are tasked with implementing, but also the underlying discourses and interests associated with state interventions in local natural resource governance in the first place. Moreover, when powerful external actors focus attention on a particular area – for example, in the context of resource appropriation – the informal practices and agreements between interface bureaucrats and communities may no longer apply.[2] Research on interface bureaucrats must incorporate and be understood in this context (Zenker, 2018).

However, the study of interface bureaucrats can help deepen our understanding of environmental governance and interventions by highlighting the everyday practices, agency and interactions that take place between state representatives and citizens. This adds a further dimension to the emphasis on domination and resistance that tends to form the main focus of critical research on environmental governance, in that it sheds light on the 'middle ground' where actors engage and reshape interventions. This can illuminate how interventionist state measures are negotiated and recast in local settings, and what happens when community members engage rather than resist such interventions.

Such a perspective should not be mistaken for naive optimism. The research in Zambia showed that, sometimes, informal arrangements between interface bureaucrats and community members do in fact result in outcomes that are better for communities than what was originally intended (Funder and Mwemba, 2019). However, engagement and negotiation with interface bureaucrats requires resources that are not available to all community members. Research on the stratified outcomes of the pragmatic practices of interface bureaucrats is thus needed. Furthermore – and importantly – the pragmatic practices and agency of interface bureaucrats can also serve to establish and extend state control through coercion and self-disciplining (Cleaver, 2018). The study of interface bureaucrats is thus as much a study of the mundane everyday ways in which a state asserts its power in everyday practice, as it is about exploring the ways in which such attempts are reshaped or negotiated.

While there is a developing literature on the informal *practices* of interface bureaucrats in natural resource governance, we know less about how their personal perceptions and social and professional relations influences their rationales, practices and agency. Here I have tentatively discussed the subjectivities and networks of interface bureaucrats in natural resource governance and sought to exemplify how this impinges on their practices. There are no doubt other entry points, and a need for engagement with other bodies of literature, such as that currently developing in organisation studies under the label of 'institutional work' (Lawrence et al., 2013), and in critical institutionalism within development studies (Cleaver and de Koning, 2015).

More broadly speaking, the study of interface bureaucrats provides an example of the need for a greater emphasis on the ways in which everyday environmental governance plays out, and how seemingly mundane actors and their practices and agency influence these processes. Such research should not replace investigation into the bigger dramas of environmental governance and politics but can substantially add to it.

Notes

1 'Rebellious norms' are mentioned in Olivier de Sardan 2014, but appear to have been abandoned in Olivier de Sardan 2015.
2 We discussed the role of central state interests and donors in Funder et al. (2018), and the limits to informal practices in Funder and Marani (2015).

References

Anders, G. (2009) 'Like chameleons: Civil servants and corruption in Malawi', in G. Blundo and P.-Y. L. Meur (eds), *The Governance of Daily Life in Africa*. Brill, Leiden.

Arakelyan, I., Moran, D. and Wreford, A. (2017) 'Climate smart agriculture: A critical review', in F. Nunan (ed.) *Making Climate Compatible Development Happen*. Routledge, London.

Barrett, S. M. (2004) 'Implementation studies: Time for a revival? Personal reflections on 20 years of implementation studies', *Public Administration,* vol. 82, no. 2, pp. 249–262.

Barrett, S. and Hill, M. (1984) 'Policy, bargaining and structure in implementation theory: Towards an integrated perspective', *Policy and Politics,* vol. 12, no. 3, pp. 219–240.

Benda-Beckmann, K. V. (1981). 'Forum shopping and shopping forums – Dispute settlement in a Minangkabau village in West Sumatra', *Journal of Legal Pluralism and Unofficial Law,* vol. 19, pp. 117–159.

Berman, B. (1992) *Control and Crisis in Colonial Kenya: The Dialectic of Domination*. East African Educational Publishers, Nairobi.

Berry, S. (1993) *No Condition is Permanent: The Social Dynamics of Agrarian Change in Sub-Saharan Africa*. University of Wisconsin Press, Madison.

Bierschenk, T. and Olivier de Sardan, J.-P. (2014) *States at Work: Dynamics of African Bureaucracies*. Brill, Leiden.

Bierschenk, T. and Olivier de Sardan, J.-P. (2019) 'How to study bureaucracies ethnographically?', *Critique of Anthropology*, vol. 39, no. 2, pp. 243–257.

Blundo, G. (2014) 'Seeing like a State Agent: The Ethnography of Reform in Senegal's Forestry Services', in T. Bierschenk and J.-P. Olivier de Sardan (eds) *States at Work: Dynamics of African Bureaucracies*. Brill, Leiden.

Blundo, G. (2015) 'The king is not a kinsman: Multiple accountabilities and practical norms in West African bureaucracies', in T. De Herdt, and J-P. Olivier de Sardan (eds) *Real Governance and Practical Norms in Sub-Saharan Africa*. Routledge, London.

Blundo, G. and Glasman, J. (2013) 'Introduction: Bureaucrats in uniform', *Sociologus,* vol. 63, no. 1–2, pp. 1–9.

Blundo, G. and Le Meur, P.-Y. (2009) *The Governance of Daily Life in Africa: Ethnographic Explorations of Public and Collective Services*. Bril, Leiden.

Bluwstein, J. and Lund, J. F. (2018) 'Territoriality by conservation in the Selous–Niassa Corridor in Tanzania', *World Development,* vol. 101, pp. 453–465.

Bourdieu, P. (1990) *The Logic of Practice*. Polity, Cambridge.

Cheeseman, N., Lynch, G. and Willis, J. (2016) 'Decentralisation in Kenya: The governance of governors', *The Journal of Modern African Studies,* vol. 54, no. 1, pp. 1–35.

Cleaver, F. (2012) *Development through Bricolage: Rethinking Institutions for Natural Resource Management*. Earthscan, London.

Cleaver, F. (2018) 'Everyday water injustice and the politics of accommodation', in R. Boelens, T. Perreault and J. Vos (eds) *Water Justice*. Cambridge University Press, Cambridge.

Cleaver, F. D. and De Koning, J. (2015) 'Furthering critical institutionalism', *International Journal of the Commons,* vol. 9, no. 1, pp. 1–18.
Cornea, N. L., Véron, R. and Zimmer, A. (2017) 'Everyday governance and urban environments: Towards a more interdisciplinary urban political ecology', *Geography Compass,* vol. 11, no. 4, e12310.
Crewett, W. (2015) 'Street-level bureaucrats at work: A municipality-level institutional analysis of community-based natural resource management implementation practice in the pasture sector of Kyrgyzstan', *Sustainability,* vol. 7, no. 3, pp. 3146–3174.
De Herdt, T. and Olivier de Sardan, J.-P. (2015) *Real Governance and Practical Norms in Sub-Saharan Africa: The Game of the Rules,* Vol. 7. Routledge, London.
De Herdt, T. and Titeca, K. (2016) 'Governance with Empty Pockets: The Education Sector in the Democratic Republic of Congo', *Development and Change,* vol. 47, no. 3, pp. 472–494.
Death, C. (2016) *The Green State in Africa.* Yale University Press, London.
Diarra, A. and Ousseini, A. (2015) 'The coping strategies of front-line health workers in the context of user fee exemptions in Niger', *BMC Health Services Research,* vol. 15, no. 3, p. S1.
Foucault, M. (2007) *Security, Territory, Population (Michel Foucault: Lectures at the College De France).* Palgrave Macmillan, New York.
Funder, M., Danielsen, F., Ngaga, Y., Nielsen, M. R. and Poulsen, M. K. (2013) 'Reshaping conservation: The social dynamics of participatory monitoring in Tanzania's community-managed forests', *Conservation and Society,* vol. 11, no. 3, pp. 218–232.
Funder, M. and Marani, M. (2015) 'Local bureaucrats as bricoleurs. The everyday implementation practices of county environment officers in rural Kenya', *International Journal of the Commons,* vol. 9, no. 1, pp. 87–106.
Funder, M. and Mweemba, C. (2019) 'Interface bureaucrats and the everyday remaking of climate interventions: Evidence from climate change adaptation in Zambia', *Global Environmental Change,* vol. 55, pp. 130–138.
Funder, M., Mweemba, C. and Nyambe, I. (2018) 'The politics of climate change adaptation in development: Authority, resource control and state intervention in rural Zambia', *The Journal of Development Studies,* vol. 54, no. 1, pp. 30–46.
Gupta, A. (2006) 'Blurred boundaries: The discourse of corruption, the culture of politics, and the imagined state', *The Anthropology of the State: A Reader,* vol. 22, no. 2, p. 211.
Hahonou, E. (2015) 'Juggling with the norms: Informal payment and everyday governance in healthcare facilities Niger', in T. De Herdt, and J-P. Olivier de Sardan (eds) *Real Governance and Practical Norms in Sub-Saharan Africa: The Game of the Rules.* Routledge, London.
Haller, T. (2012) *The Contested Floodplain: Institutional Change of the Commons in the Kafue Flats, Zambia.* Lexington Books, Lanham, MD.
Hamani, O. (2014) '"We make do and keep going!" Inventive Practices and Ordered Informality in the Functioning of the District Courts in Niamey and Zinder (Niger)', in T. Bierschenk and J.-P. Olivier de Sardan (eds) *States at Work. Dynamics of African Bureaucracies.* Brill, Leiden.
Hupe, P. and Hill, M. (2007) 'Street-Level bureaucracy and public accountability', *Public Administration,* vol. 85, no. 2, pp. 279–299.
Kairu, A., Upton, C., Huxham, M., Kotut, K., Mbeche, R. and Kairo, J. (2018) 'From Shiny Shoes to Muddy Reality: Understanding How Meso-State Actors Negotiate the Implementation Gap in Participatory Forest Management', *Society and Natural Resources,* vol. 31, no. 1, pp. 74–88.

Lawrence, T. B., Leca, B. and Zilber, T. B. (2013) 'Institutional work: Current research, new directions and overlooked issues', *Organization Studies,* vol. 34, no. 8, pp. 1023–1033.

Le Marcis, F. and Grard, J. (2015) 'Ethnography of everyday ethics in a South African medical ward', in T. De Herdt and J.-P. Olivier de Sardan (eds) *Real Governance and Practical Norms in Sub-Saharan Africa: The Game of the Rules.* Routledge, London.

Li, T. M. (2005) 'Beyond "the state" and failed schemes', *American Anthropologist,* vol. 107, no. 3, pp. 383–394.

Li, T. M. (2007a) 'Practices of assemblage and community forest management', *Economy and Society,* vol. 36, no. 2, pp. 263–293.

Li, T. M. (2007b) *The Will to Improve: Governmentality, Development, and the Practice of Politics.* Duke University Press, Durham, NC.

Lind-Riehl, J. F., Mayer, A. L., Wellstead, A. M. and Gailing, O. (2016) 'Hybridization, agency discretion, and implementation of the US Endangered Species Act', *Conservation Biology,* vol. 30, no. 6, pp. 1288–1296.

Lindegaard, L. S. (2018) 'Adaptation as a political arena: Interrogating sedentarization as climate change adaptation in Central Vietnam', *Global Environmental Change,* vol. 49, pp. 166–174.

Lipsky, M. (2010) *Street-Level Bureaucracy: Dilemmas of the Individual in Public Services* (Expanded edition). Russell Stage Foundation, New York.

Long, N. (1989) 'Encounters at the Interface'. Agricultural University of Wageningen, Netherlands.

Lund, C. (2006) 'Twilight institutions: Public authority and local politics in Africa', *Development and Change,* vol. 37, no. 4, pp. 685–705.

Lund, C. and Boone, C. (2013) 'Introduction: Land politics in Africa–constituting authority over territory, property and persons', *Africa,* vol. 83, no. 1, pp. 1–13.

Maier, C. and Winkel, G. (2017) 'Implementing nature conservation through integrated forest management: A street-level bureaucracy perspective on the German public forest sector', *Forest Policy and Economics,* vol. 82, pp. 14–29.

Mosse, D. (2005) *Cultivating Development: An Ethnography of Aid Policy and Practice.* Pluto Press, London.

Nuijten, M. (2005) 'Power in practice: A force field approach to power in natural resource management', *Journal of Transdisciplinary Environmental Studies,* vol. 4, no. 2, pp. 3–14.

Olivier de Sardan, J.-P. (1999) 'A moral economy of corruption in Africa?', *The Journal of Modern African Studies,* vol. 37, no. 1, pp. 25–52.

Olivier de Sardan, J.-P. (2014) 'The Delivery State in Africa. Interface Bureaucrats, Professional Cultures and the Bureaucratic Mode of Governance', in T. Bierschenk and J.-P. Olivier de Sardan (eds) *States at Work: Dynamics of African Bureaucracies.* Brill, Leiden.

Olivier de Sardan, J.-P. (2015) 'Practical norms: Informal regulations within public bureaucracies (in Africa and beyond)', in T. De Herdt and J.-P. Olivier de Sardan (eds) *Real Governance and Practical Norms in Sub-Saharan Africa.* Routledge, London.

Ranger, T. O. and Hobsbawm, E. J. (1983) *The Invention of Tradition.* Cambridge University Press, Cambridge.

Ribot, J. (2007) 'Representation, Citizenship and the Public Domain in Democratic Decentralization', *Development,* vol. 50, no. 1, pp. 43–49.

Ricks, J. I. (2016) 'Building participatory organizations for common pool resource management: Water user group promotion in Indonesia', *World Development,* vol. 77, pp. 34–47.

Sareen, S. (2016) 'Seeing development as security: Constructing top-down authority and inequitable access in Jharkhand', *South Asia Multidisciplinary Academic Journal*, no. 13.

Scott, J. C. (1985) *Weapons of the Weak: Everyday Forms of Peasant Resistance*. Yale University Press, New Haven, CT.

Scott, J. C. (1998) *Seeing Like a State: How Certain Schemes to Improve the Human Condition Have Failed*. Yale University Press, London.

Sevä, M. and Jagers, S. C. (2013) 'Inspecting environmental management from within: The role of street-level bureaucrats in environmental policy implementation', *Journal of Environmental Management*, vol. 128, pp. 1060–1070.

Sikor, T. and Lund, C. (2009) 'Access and property: A question of power and authority', *Development and Change*, vol. 40, no. 1, pp. 1–22.

Stepputat, F. and Nuijten, M. (2018) 'Anthropology and the enigma of the state', in H. Wydra and B. Thomassen (eds) *Handbook of Political Anthropology*. Edward Elgar Publishing, Cheltenham.

Taylor, M. (2018) 'Climate-smart agriculture: What is it good for?', *The Journal of Peasant Studies*, vol. 45, no. 1, pp. 89–107.

Telle, K. (2014) 'Policing and the politics of protection on Lombok, Indonesia', in P. Albrecht and H. M. Kyed (eds) *Policing and the Politics of Order Making*. Routledge, London.

Therkildsen, O. (2014) 'Working in Neopatrimonial Settings: Public Sector Staff Perceptions in Tanzania and Uganda', in T. Bierschenk and J.-P. Olivier de Sardan (eds) *States at Work: Dynamics of African Bureaucracies*. Brill, Leiden.

Vandergeest, P. and Peluso, N. L. (1995) 'Territorialization and state power in Thailand', *Theory and Society*, vol. 24, no. 3, pp. 385–426.

Weatherley, R. and Lipsky, M. (1977) 'Street-level bureaucrats and institutional innovation: Implementing special-education reform', *Harvard Educational Review*, vol. 47, no. 2, pp. 171–197.

Weick, K. E. (1995) *Sensemaking in Organizations*, Vol. 3. Sage, Thousand Oaks, CA.

Weilenmann, M. (2017) 'Project law: Normative orders of bilateral development cooperation and social change: A case study from the German agency for technical cooperation', in F. von Benda-Beckmann and K. von Benda-Beckmann (eds) *Mobile People, Mobile Law: Expanding Legal Relations in a Contracting World*. Routledge, London.

Whitfield, S. (2016) *Adapting to Climate Uncertainty in African Agriculture: Narratives and Knowledge Politics*. Routledge, London.

Zenker, O. (2018) 'Bush-level bureaucrats in South African land restitution: Implementing state law under chiefly rule', in O. Zenker and M. V. Hoehne (eds) *The State and the Paradox of Customary Law in Africa*. Routledge, London.

Zietsma, C. and Lawrence, T. B. (2010) 'Institutional work in the transformation of an organizational field: The interplay of boundary work and practice work', *Administrative Science Quarterly*, vol. 55, no. 2, pp. 189–221

10

INSTITUTIONAL RECOGNITION POLITICS IN FORESTRY

When the plurality of state law undermines local democracy

Papa Faye

Introduction

Legal pluralism – 'the presence in a social field of more than one legal order' (Griffiths, 1986: 1)[1] – and the plurality of institutional bodies or organizations, or, institutional pluralism, intersect in practice. However, there are at least two limits to a better understanding of legal pluralism. The first limit is that legal pluralism is not only the imbrication of different orders of law in a social field (Benda-Beckmann, 1988; Merry, 1988) but can also be internal to the same domain of law (i.e., state legal order). The state legal order is itself internally multilayered and plural, since it is composed of often-contradictory legal instruments conceived at the various administrative and technical levels, from the national to the local, serving their particular interests, which are legitimized through their diverse legal interpretations, or 'forum shopping' (Benda-Beckmann, 1981: 117). This form of legal pluralism can be referred to as internal legal pluralism, as opposed to classic legal pluralism, which refers to the coexistence of different orders of law in a given social field (Benda-Beckmann, 1988; Griffiths, 1986; Merry, 1988).

The second limit is that the intersection between legal pluralism (no matter how it is defined) and institutional pluralism and their consequences on politics in general and democracy in particular are not well-known. Institutional pluralism originates from the practices of intervening actors such as state, development or environmental projects and programs of forming new or empowering existing local institutional bodies rather than others – referred to as 'institutional recognition' (Ribot et al., 2008). Similar interventions are very common in the area of natural resource governance, particularly in forestry where the mobilization of local people has become a plea since the 1992 Rio World Summit. This chapter contends that state law is not always a single legal domain; it is usually a multilayered order of law composed of many regulations which operate as separate legal domains, regardless of the normal hierarchy of state law. Therefore, I argue that, because of the plurality of institutional bodies without clearly defined relations of accountability as a consequence of the politics of recognition of intervening actors, there is often a form of legal pluralism even within a domain of law although this one is believed to be a single legal domain. In this chapter, I analyze this internal pluralism and narrate the intersection between legal and institutional pluralisms and their effects on democracy in decentralization contexts.

Decentralization policy emerged in the developing world in the 1990s (Ribot, 2002). It was expanded to the sector of natural resources including forest governance during the same decade (Faye, 2006; Gellar, 1997; Moore et al., 2000). In Senegal, many sectors of power were transferred from the state to elected local governments since the 1996 Decentralization Act, expanded by the 2013 Decentralization Act, which recognized the municipalities and Departments (forms of provinces) as administratively autonomous elected local governments. Theoretically, nine domains of power are now in the hands of these local governments[2] including the authority over the environment and renewable natural resources in general and in particular non-gazetted forests within their territorial jurisdictions. So to put decentralization policy into reality, in each targeted sector, projects were designed and supported either by the government or international donors. In the forest sector, the World Bank, USAID and GIZ have been the most known and long-standing funders of the implementation of decentralization policy. The main reason why the government called upon donors is that the forest department argued that the costs of the expansion of participatory forest management plans (FMPs) required international funding (Faye, 2015a, 2015b). However, for purposes of delivering quantitative outputs rather than procedural outcomes, projects choose to partner with non-elected local actors or create new ones in the name of participation (Faye, 2018; Ribot et al., 2008) and 'depolitization' (Faye, 2015a, 2015b, 2018). By recognizing new institutional actors in the local arena, these projects further institutional pluralism in their areas of intervention.

In Senegal, forestry interventions are carried out in collaboration with the forest department. They have created participatory organizations called committees that act in their own domain in isolation from the elected local governments, the lawful institution they should have partnered with according to the spirit of decentralization policy. In this context, I use 'institutional pluralism' to refer to the coexistence of multiple organizations, including village chiefs (who try to reinvent themselves in new practices of forestry) and municipal councils. The newly created project committees operate as separate bodies of enforcement of the FMPs, which are really a law *per se* because they define rules and generate organizations for their own legal enforcement. Indeed, in some instances, those FMPs at the very local level contradict some aspects of state and sectorial legislation, or are contradicted by more practical regulatory instruments such as ministerial orders and circulars. In this respect, projects further legal and institutional pluralism at the same time.

The next section reviews the literature and theories of pluralism. The following presents the case studies and the methods. The subsequent section describes the legal pluralism found in the study areas, and its internal aspects. The chapter then demonstrates how, by crafting new participatory committees, projects have intensified the rural institutional landscape after decentralization. The features of the intersection between legal and institutional pluralism are then described. The final section draws out and discusses the main results and their implications for local democracy in the context of decentralization reform (here, with special focus on decentralized natural resource management).

A review of literature and theories of pluralism

This section describes different theories of pluralism and similar concepts in the disciplines of social sciences and beyond, emphasizing how our conception of pluralism fits into that discussion. It also shows the ways in which conceptions of pluralism have also been evolving, adapting to the complex and dynamic evolution of democratic processes across the world.

Pluralism[3] is reified in Western thought and later in Western-driven development interventions in the Global South, as either a feature of or condition for (liberal) democracy. Dahl[4] and Lindblom (1976: 277) viewed 'polyarchy' – a multiplicity of many powerful groups in which no one group dominates the others – as a pro-democratic political system. For political scientists of the United States, 'polyarchy' in the 1960–1970s characterized the end of the rule of businessmen in American cities. Among its most democratic achievements were: the domestication of power and reduction of coercion; the promotion of citizens' consent in the long run; and the fostering of peaceful benefits to all the claimants (Dahl cited in Manley, 1983: 369). More recently, Ostrom (1999) advocated that 'polycentricity' of government introduced a 'balance of power' between the higher level and the lower.

Anthropologists also study pluralism, mainly focusing on the coexistence of multiple laws (legal pluralism) and, in development anthropology, authorities and laws at the same time. Legal anthropologists have usually studied the intertwining of folk and state law (Griffiths, 1986; Merry, 1988), or the merging of both 'within the same process, or in aggregate processes' (Benda-Beckmann, 1988: 899). There also exists a multilayeredness of laws in a single legal order rather than in coexisting different legal orders, as usually studied in legal anthropology. I argue that different state laws operate as separate domains of law within a single sector as a product of the discretionary regulatory power held by public civil servants at various state structures and scales.

Some development anthropologists opt for a 'dual pluralism' that embraces the plurality of institutions (bodies exercising power) and of norms and regulations. 'Polycephaly'[5] (*polycéphalie*), as conceptualized by Bierschenk (1995: 8), and later by Bierschenk and Olivier de Sardan, refers to 'a myriad of institutions and decision-making loci (traditional, neo-traditional, informal and formal), which do not have strictly defined competences and lack clearly organized hierarchical relations of subordination and dependence' (2003: 166).

My perspective shares two elements with those of development anthropologists: the study of pluralism as a phenomenon whose complexity has increased with decentralization reform, and the dual understanding of pluralism as the coexistence of actors and of overlapping repertoires of regulation. However, I will break with them, for analytical and methodological reasons, by differentiating legal pluralism (herein the plurality of legal orders, or plurality of laws within a single order) from institutional pluralism (the plurality of institutional and organizational actors), to better explain their intersection in practice.

These development anthropologists did not emphasize, as this chapter does, the regulatory power of the many institutional bodies involved in a given sector or arena and the fact that the regulatory instruments they produce operate as separate legal domains within state law. It is important to demonstrate that it is the power of these actors to craft new regulations that pushes them to produce competing 'new rules of the games.' Technical civil servants such as forest agents may use their power to issue regulations to maintain and sustain their authority over forest resources and related decision-making processes. Usually, as seen in this chapter, while institutional actors at the higher-level of government may produce legal instruments that support democracy formation and consolidation, lower-level ones may invent new regulatory tools to shrink this window so as to maintain control over resources and the privileges behind that.

Institutional pluralism as defined in this chapter may not be very different from meanings of multilevel governance – the principle that vertical and horizontal interactions are needed among the multiple administrative levels at which actors involved in governance are located (Forsyth, 2009; Nunan, 2018). This chapter looks at the legal effects of multilevel governance at each of the administrative levels involved in decentralized forest governance regulatory instruments that operate in autonomous domains of law within the same state legal and institutional orders.

Case studies and methods

This section presents the projects studied and the sites where the material was collected. It also describes the methodological approaches and instruments used during fieldwork and to analyze the data collected.

Following the 1996 decentralization reform, PROGEDE was the leading participatory project devised for forest management and commercial exploitation of forest resources. It is also the most long-standing as it is still under implementation, operating an extension of its second phase. PROGEDE started effectively in 1998 and was devised by the forest department and the directorate of energy. It has two components, responsible, respectively, for fuelwood supply and energy demand. I focused on the supply component, which was headquartered in the forest department. The main objective of the supply component is to promote sustainable fuelwood supply while ensuring 'rational' forest management and transferring the power of management to the local people residing near the forests. PROGEDE had two phases, and between them a transition period and a temporary cessation of activity. PROGEDE-I ran from 1998 to 2004. The transition period lasted from 2005 to 2008. The project ceased activity from 2009 to 2011 before PROGEDE-II returned in 2011, with a mandate to continue operation through to 2015.

PROGEDE-I and II received funding from various donors, as well as support from the Senegalese government, though the main donor was the World Bank, which was also responsible for monitoring the project. PROGEDE-I had two regional offices led by foresters. Both offices were supervised by a national coordinator working under the authority of the director of the forest department. In PROGEDE-II, national coordination remained with the forest department, but regional offices were handed over to independent experts rather than foresters. However, PROGEDE-II offices were dependent on regional forest service offices.

The case material presented here was collected in the commune of Missirah, whose community forests have been targets of the two management projects. A community forest is any forest that is not gazetted in the state forestry domain and is located within the territory of a rural Commune. So the municipal council and the mayors that lead them gain official management power over community forests from the Forest Act No. 98–164 of 1998, Art. R9. This has been expanded by the Forest Act No. 2018–25 of 12 November 2018 and operational Order No. 2019–110 of 16 January 2019. The study took place within the department and region of Tambacounda and focused on the community forest managed by PROGEDE and known as Missirah-Kothiary managed forest.

Methods were mainly open interviews (formal and informal), semi-structured interviews and participant observation. Fieldwork periods spanned from 2004 to 2013 with more intensive fieldwork stays occurring, respectively, from April 2004 to May 2006, March 2012 to August 2012, November 2012 to May 2013, and from December 2013 to January 2014. There were additional on and off field trips from 2017 to 2018, enabled by a grant I earned from the Open Society Foundations Fellowship Program in New York. The chosen ethnographic methods enabled the gathering of accounts about certain central tensions; including competition among the existing institutions and the rationales they mobilize to justify their deeds and positions.

To understand the rationales from above, I 'studied up' (Nader, 1972), interviewing high-level officials to understand some of the dynamics being played out at lower levels. To identify the contradictions or conflicts between the higher- and lower-level perceptions and practices, I went back and forth on the ground using the ethnographic methods cited above. Most of the interviews were recorded, except where the interviewees wished not to be recorded or where the issues discussed were sensitive, and then transcribed for processes of data coding and analysis.

By doing this 'studying through' (Wright and Reinhold, 2011), I could identify areas of dispute and examine the conflicting rationales that implicated specific laws. To assess the gaps or differences in the interpretations of particular laws by the different actors, I did a systematic content analysis of standard documents and all legal instruments concerned with forest management. The informal norms of local people, since these are not inscribed in legal documents, emerged from the ethnographic work. Additionally, I borrowed from new institutionalism in (economic) anthropology to explain the linkages between discourses, laws and practices and to understand how the interactions between these also shaped the current institutional landscape (Ensminger, 1992; extended by Haller, 2010).

The two forms of legal pluralism

This section shows the twofold character of legal pluralism in Senegalese forestry, focusing on legislation and regulations concerning charcoal, which is currently the most disputed commercial forest resource. The first subsection describes the overlapping of different legal orders at the decentralized level – the classic legal pluralism. The second portrays internal legal pluralism: the multilayered interests within state structures that are operating as separate domains of law.

The classic legal pluralism or resistance to legal centralism

A study of classic legal pluralism in Tambacounda reveals an encounter among diverse forms of law. The precolonial laws resisted the extractive colonial legal overlay, as well as post-colonial domination. However, in practice all the laws are overlapping and being 'shopped' to serve particular claims and interests.

In the precolonial period, forests were part of the patrimony of geographically delineated people and locations. The case of ethnic groups in the study area illustrates this. The dominant ethnic groups are Fula and Mandinka. The Fula used to have a rationale of utility towards forest, grazing, harvesting trees for their livestock, and so on, whereas Mandinka entertained sacred relations with forests. Of course, non-sacred forests were subject to domestic usages such harvesting wood for habitat construction and collecting wood for cooking as well as wild fruits and honey collection. But, in general, the forests were intimately linked to Mandinka's sociopolitical system, as the rule of new kings was celebrated in the sacred forest where the supernatural device (tree, forest or wild animal) called *jalan* was always placed (Niane, 1989). Each village and kingdom had its *jalan* for its own protection and cult. Therefore, the Mandinka ethnic group had precolonial organizations dedicated to forest surveillance and protection. They took diverse names such as *tesito* (means voluntary work for collective interests), *wula nafaa* (means rich or forests with high value) and *wulo demaalaa* (forest protection or surveillance).

Even the Fula group, which seemed to have a more utilitarian relationship to the forests, had rules and regulations to control access to forests for grazing. Territorialization of forest resources, belonging to the community and social networks, and rights to natural pools were the main institutions; all were under the authority of the village chief known as *jomoo wuro* (village owner in Fula).

During the colonial period, all the forests fell under the control of the colonizing State, France. The development of railways valued the forests as fuel for the trains (Ribot, 2001a). As the region of Tambacounda grew around and along the railways linking Dakar (Senegal) to Kayes (Mali), the majority of its forests were gazetted (referred to as 'classified' in French) for exclusive control of the colonial state. For example, the forests of Tamba Nord (75,000 ha) and of Ouli (14,000 ha) were gazetted, respectively, in 1939 and 1943.

For forests open to commercial exploitation, management aimed at providing French industries with raw materials, wood in particular. Forest management was extractive and dictatorial (Gellar, 1997; Kanté, 2006). Rural people were discriminated against while urban populations were favored (Ribot, 2001a). The urban areas of Dakar, Gorée, Rufisque and Saint Louis were classified as 'communes' under French rule and were privileged with French citizenship. Among the French citizens, those originating from France were favored over the 'indigenous French' (ibid.).

After independence, the Senegalese State continued this logic of dictatorship and exclusion of rural people from the management of forest lands and resources. On 17 June 1964, the National Domain Law (No. 64-46) was promulgated. It denied any property rights on land for people without private title to their plots at the date of promulgation, no matter how long they had been using the land, or what rights they were claiming. Only those with private titles or valuable investment (i.e., planted trees and equipment) could claim to own the land as private property. This investment was called *mise en valeur.* Therefore, more than 95% of the whole national territory fell under the control of the state, including forest spaces. The law meant that rural people did not 'own' their agricultural and grazing lands, including immediate forests; they could retain use rights to those lands, but even use rights could be lost if the land became earmarked for public interest projects.[6]

In 1972, the Senegalese government undertook administrative and territorial reform. From that reform, in addition to urban communes, new public entities emerged at the rural level. They were known in French as *Communautés rurales*. They were shifted into communes following the promulgation of the third Decentralization Act No.2013-10 of 28 December 2013. But for simplification and to avoid confusion with urban communes that preceded them, I call them herein 'rural communes.' The rural communes were governed by elected local governments but were still headed by sub-prefects, state-appointed civil servants. Sub-prefects had therefore a double legitimacy, as they also acted as heads of the *arrondissements*[7] and the rural communes within it. The power to distribute use rights (and to withdraw use rights) among the rural people living within rural communes was transferred to the rural municipal councils.

However, this reform did not abrogate the 1964 National Domain Law, it just decentralized the level of land-use decision-making from the national to a lower level. The status of forestry spaces remained the same; they were mainly part of the National Domain, as were state-gazetted forests like parks and any other protected forest block. Rural people were barred from the internal delegation of use rights (lending, rent or sale, legacy), although in practice use rights were and are still being delegated within the villages, families and from family to family or, say, among the rural people (Faye et al., 2012).

In 1996, political decentralization reform, referred to as 'improved decentralization,' was launched. It created new levels of governance (the regions, through 'regionalization') and reorganized the urban communes. For the first time, the regions were to have, together with the governors (who were government delegates), democratically elected authorities similar to those of urban and rural communes. Since 1996, the state political architecture has been composed of three local entities that are said to autonomously administer those public affairs that are within their competence. The power to manage renewable natural resources, excluding non-state and private forests, is one of the nine powers transferred to the rural communes since the 1996 decentralization reform.

This legal evolution led to the promulgation, in 1998, of the so-called Decentralizing Forestry Code. The Code's main innovation was the creation of 'community forests' and the transfer of their management to the rural communes (I focus on rural forests herein) where they were located. Theoretically, as the National Domain law was not abrogated by the Decentralized Forestry Code, community forests remained part of the national domain.

However, the community forest has to be fully managed according to local conditions, priorities and wants, as long as decisions are consistent with the existing forestry laws (Law No. 96-06: art. L195, L200; Law No. 96–07: art. R45–6). Unlike state forests, the community forest is registered for the profit of neither the state nor the rural municipal council. Moreover, the rural municipal council cannot have a private domain of land, as it is barred from land ownership; the rural municipal council can only manage and (re)allocate land among the rural peoples for agricultural purposes.

Legal pluralism in Senegal's forestry is apparent in the use of community forests. Indeed, the forest department, which stands and acts for the state in the forest sector, contends in practice that a rural municipal council is the 'manager of the community forests but does not own it.' Drawing on the 1964 National Domain Law, foresters claim that forests are 'national goods' that should be managed in ways that profit all legal Senegalese citizens.

Moreover, because of that law, the forest department is still exercising the power to monitor and prevent illegal forestry activities but to confer commercial access rights: exploitation and hunting permits and the royalties arising from these (Law No. 98–164: art. R20). Decentralization reform has also not changed forestry taxation; it just gives a share to rural municipal councils on the returns collected for infractions committed within their jurisdiction. In such cases, the rural municipal council's share is 70 percent of the total amount collected. In practice, however, the rural municipal councils rarely fully receive this money from the forest service.

Rural people's perspective on forestry rights differ, however; they continue to believe that they 'own' the lands and forests nearby their villages. The presence and precedence of the forest department in forest issues, legitimized by statute, has not eliminated the customary institutions. At the beginning of the forestry projects studied here, many villages had still until 2005 maintained those customary institutions (Faye, 2006). Under the authority of village chiefs (descendants of the first dwellers but also state non-salaried appointees), they continued policing forests, using violence when external woodcutters licensed by the forest department showed resistance.

This was the case in 1994 in Sabi Mulesi when the leaders of the customary forestry policing institutions were jailed on charges of inflicting deliberate injury on woodcutters who were operating within their village forest. According to state laws, those woodcutters were acting legally, because they had been hired and registered by Senegalese licensed forest merchants. Village laws, however, forbade any form of green wood harvesting, except for domestic uses like home-building. While no longer recognized nationally, customary rules were still being exercised locally.

In both ethnic groups, classic legal pluralism manifests through the perception that forests surrounding a village belong to that village and are called 'village forests.' People refer to forests by associating them with the villages to which they belong to according to local views and norms; for example by saying, in Fulani and Mandika respectively, *ladde Barkeyel* or *wulo Diyabugu*, 'the forest of the village of Barkeyel or Diyabugu.' A village forest consists of forest spaces that are close to that village, just beyond the village agricultural domain.

As for village forest, inter-village forests are referred to in association to villages that share it. For instance, *ladde Barkeyel e Gurel Bokar* means 'the forest of the villages of Barkeyel and Gurel Bokar.' Significantly, a forest is locally perceived as the collective property of a village or villages, depending on the situation of the forest vis-à-vis the villages. These perceptions confer to the village chiefs, in the name of customs and traditions, the capacity to control access to village forests. For example, any mobile herders that want to settle within the area around a village must be hosted by a village resident and presented to the village chief for acceptance. Ownership of 'far-off' forests was not claimed at all by any village, and access to these obeyed the rule of 'first come, first served' (Bromley, 1991; Idelman, 2008).

Internal legal pluralism

Legal pluralism, as in overlapping jurisdictions, is apparent even among state laws. The legal order in Senegal does not create consistency among legislative and regulatory levels. Different layers of interests and practice serve as separate legal domains within the State.

The General Decentralization Law contradicts the 1964 National Domain Law as to the level and the author of certain kinds of decision-making. According to the National Domain Law, decisions on forests can only be taken at the national level, by the technical arm of the state, which is the forest department. According to the decentralization law, however, this decision is made at the rural municipal council's discretion; the rural municipal council can solicit the forest services or hire foresters, for technical support where needed. The so-called Decentralized Forestry Code also reduced the scope of rural municipal councils' governance by introducing a technical and also regulatory instrument (referred to as the FMP) as a precondition for the effective transfer of community forest management power to rural municipal councils. In practice, the FMPs predetermine the types of use acceptable in a particular forest parcel, choose and empower actors, and, as far as commercial resources are concerned, organize the exploitation in time and space.

The implication of predetermining those elements is that the elected local governments have lost the most important discretionary power they gained from decentralization reform: the power to grant a required permission before any woodcutting can take place in community forests. The invention of the FMP requirement, and its expansion throughout the regions where forests suitable for charcoal are located, were initiated by the forest department. Interestingly, the forest department is the state body that had to negotiate or put pressure on the Rural Commune Mayors, which used to resist giving permission for charcoal production within the community forest of their rural commune (see Kanté, 2006; Faye, 2006; Ribot and Faye, 2007). Since a FMP subtly imposes charcoal, withdrawing the management power from the rural commune mayors and their mayors, it reduces the high transaction costs associated with gaining permission annually.

Mayors of rural communes were expecting their rights to continue and did not know they had lost them once they approved the FMP. The rural commune mayors did not know that their power had been taken away because its withdrawal was not explicitly inscribed in any legal instrument, not even in the FMPs.

Furthermore, in 2009, the forest department issued a circular to officially withdraw from rural municipal councils the 'veto' power on charcoal. Circular no. 000209, dated 29 January 2009 says, 'Once they have approved the FMP for their forest, the rural municipal councils give up the right to deliver permission for charcoal production.' In this manner, the initial power devolved to rural municipal councils through Decentralization and Forestry Acts was taken away through the FMP and the regulatory instruments such as this circular meant to implement the FMP on the ground. Therefore, with the establishment of the FMP, the concerned rural municipal council is left at the edge of decision-making processes around the use of forests. Their role became restricted to certifying decisions already taken by public forest agents during annual meetings held at the departmental council level to allocate use rights among local users and external private merchants (see below for more on this).

In addition to circulars, ministerial orders also serve particular interests within the state, and contribute to contradictions among the state legal instruments. Under the National Domain Law, the ministry of environment, through the forest department, was annually issuing a ministerial order to set the timeline and specific conditions for the exploitation of commercial forest resources. That ministerial order still organizes commercial exploitation despite the power transfer implied by decentralization policies, and despite the FMP that are said to have been put in place for the purpose of favoring the sustainable exploitation of resources. That ministerial order serves to bring back state control over forests and over revenues generated by taxes levied on forest resource extraction, despite the decentralization reform that is promoted by law.

More locally, at the regional forest service offices, some regulations (many of which are not even written down) also operate separately. For instance, in 2010, the then head of the Tambacounda office decided to punish local people who stopped installing firebreaks along the charcoal-production forest parcels. The creation of firebreaks as a method of fire prevention was widespread during the implementation of FMPs, and villagers were assigned the painful task of building them.

However, since their mobilization was dependent to the material incentives they were receiving from PROGEDE-I, villagers just stopped doing those activities when the compensation ceased. As they disengaged from cutting firebreaks, the head of the regional forest office informally decided to hand over the job to four of his local allies, in exchange for privileges that would help them in the charcoal business. Note that these four men were not really doing the job by themselves, but were *hiring villagers* to cut firebreaks. While the firebreaks were being cut (a process that can take more than four months), rural people were denied access to those parcels, as well as the right to produce charcoal from them.

That decision contradicted the ministry's annual order, which called for the official start of charcoal production as soon as the ministerial order had been publicized. It also contradicted the FMPs presented by the forest department, and the idea that management projects would open up the charcoal chain to the villagers for rural poverty alleviation and sustainability. However, that decision served to keep the permits due to villagers in forest agents' hands instead; foresters then sold the permits on the black market to private forest merchants, and were also accused by other villagers and some project officials of receiving bribes from the privileged four persons who were also hiring those villagers for the cutting of firebreaks.

This section has shown that each of the legislative and regulatory bodies uses different elements of regulation to enforce particular interests, and that many of these elements contradict each other. It has also demonstrated that the legal interventions by state structures are not consistent, nor are their interests; and both rest on different interpretations of the law as well as on the legal competence of the institutions involved. Legal pluralism therefore is both an external and internal practice at the same time.

The following section reveals how the so-called 'participatory' management projects create yet another layer of organizational actors in the already plural institutional landscape.

Project politics of recognition and reification of institutional pluralism

This section demonstrates how the management projects launched after the 1996 decentralization reform, such as PROGEDE, have aggravated the complexity of the rural institutional landscape, creating parallel committees in the name of participation. It uses the case of PROGEDE to describe the institutional choices made and how this resulted in institutional pluralism without clearly defining relations of accountability among the coexisting organizations.

PROGEDE supported a wide range of FMPs and, while pushing out the rural municipal councils, opted for the creation of village-based organizations to implement them. However, PROGEDE's organizational scheme differed slightly in the project's two phases (PROGEDE-I and PROGEDE-II).

In each nearby village of the managed forest, PROGEDE-I created a committee called the village committee and all these committees were regrouped in an inter village committee composed by two delegates from each village committee. Note that there were intermediary committees in-between the village committees and the inter-village committee known as Block Management Committees. Because local people contested the technical delineation of the forest blocks that then dictated the block management committees, these intermediary committees did not function at all. Villagers disagreed about the placement of villages with no socio-historical affinities in one forest block, and the isolation of some villages from others with which they have traditional affinities, including shared use of community forest spaces.

Village chiefs had a significant influence on the constitution of the village committees; they led the process under the supervision of PROGEDE-I's facilitators. Although democratic election of village committee members was officially envisaged by the project, village chiefs either designated themselves or designated their own sons, nephews and lower caste people (Faye, 2006) to the committees. The designation of such socially subordinate people indirectly guaranteed the village chief's control of access to the forests, which they had formerly exerted in the customary period. In villages where village chiefs failed to designate committee leaders, conflicts arose. This was the case in Diyabugu, between the Imam who led the committee and the village chief. The village chief accused the Imam of replacing him as the main authority in the village and said he would not give up without a fight. They stopped talking to each other, engaging in a contest to gather supporters both from the village side and the project.

PROGEDE's rationale for the creation of village committees revolved mainly around the execution of project activities and the implementation of the technical prescriptions imposed by the forest department for the sake of resource sustainability. Foresters interviewed contended that in a context of insufficient foresters and lack of equipment, and the impossibility of mobilizing rural municipal council members for the project activities, only nearby villagers, if well organized, could mobilize enough to protect the forests. So the role of the village committee was to implement the technical prescriptions contained in the FMP, for example, with cheap labor or 'participatory corvée' – 'participatory but forced work' (Ribot, 2001b: 328). Another official motivation was that the creation of village-based committees and their empowerment meant that power transfer would profit villagers directly (*Lebu Bi*, 06 July 2012) rather than the 'politicized' rural municipal councils – the idea that rural municipal councilors would act not for the benefit of all villagers but on behalf of the political parties to which they belong.

The inter-village committee formally replaced the rural municipal council in the management of forest, and charcoal in particular. Thus, rural commune mayors disliked PROGEDE-I's participatory organizational scheme, which distanced and isolated them from their constituents. Therefore, during the temporary cessation of PROGEDE, they dissolved the inter-village committees and crafted new inter-village committees composed of delegates from the rural municipal council and delegates from the forest blocks. Rural commune mayors borrowed from Wula Nafaa's idea of creating spatially based organizations around the forest blocks rather than residentially based institutions such as village committees.

When launched in 2011, PROGEDE-II abandoned the approach of using village committees and opted for a college of delegates from forest blocks, creating a private association for the management of the community forests. The logic of the forest department and PROGEDE (note that the project is headed by the forest department) was concerned with efficiency and sustainability of actions. They drew lessons from the disengagement of villagers from village committee interventions when PROGEDE-I temporally ceased activity in 2008, and decided that a private institution would work better and be long-lasting.

They also argued for more transparency in the relations between the new association and the rural municipal councils, to whom the management power legally belongs, and in the management of the funds derived from the local taxes levied on charcoal. Increased transparency had been a demand from rural municipal councils during PROGEDE-I, and the lack of transparency was among the reasons for the rural municipal council's dissolution of the project committees during PROGEDE's temporary retreat. Therefore, PROGEDE-II decided to devise two kinds of functions: management and auditing. While the association manages forest-access rights, the rural municipal council was to audit the management of the cash being generated from local charcoal taxation. On the advice of a consultant, PROGEDE-II also planned for the payment of the association members who would be at the forefront of the groundwork.

So efficiency, sustainability of activity and transparency were the principles of PROGEDE and the forest department's official rationale for forming associations. However, the real rationale for creating a private organization for the management of a public power was to legally bar rural municipal councils from dissolving the project's institutions at the end of the intervention as they had done in PROGEDE-I (*Mbaati,* a PROGEDE-II official, 08 March 2013). As a private body, an association can only be dissolved by the state structures that have capacity to create it. But, as one Rural Commune Mayor said, 'we cannot dissolve their association, but we can stop delegating our power at any time' (*Alaaji*, 26 December 2012).

The process of forming the association started at the village level, where four persons were chosen to constitute the 'contact group.' A contact group is composed of two men and two women, including two young persons. The contact groups of the villages within each forest block designated five delegates. The five delegates from each of the five areas, together, formed an Executive Committee of 25 delegates. From that Executive Committee, six persons were chosen to compose the Executive Bureau. Because the Executive Bureau's role is focused on administrative and financial tasks, the delegates of each area appointed a guard (*Surveillant de bloc*). The guard is responsible for organizing intervention against fires, collecting taxes, checking the legality and regularity of actions in the forest and denunciation of smugglers and technical prescription breakers.

The rural municipal councils view the association as an attempt by PROGEDE-II and the forest department to again expel them from forest governance. Indeed, the mayors of the rural commune of Missirah at that time never approved the formation of the association: he refused to attend the launch assembly meeting but gave permission for the setting of a renewable six-month trial contract for the association (Participant observation at the assembly on 30 November 2012; interview with a PROGEDE-II officer). Because forest management has been, since 1996, a legal power of the rural municipal council, mayors now make sure that their signature is obtained before any organization is designated to work in that domain. In PROGEDE-I, at the inception of decentralization reform, Rural Commune Mayors were not requiring organizations to obtain their signature. However, after a decade of decentralization and project interventions, in PROGEDE-II they began demanding respect for the rule of law.

If in PROGEDE-I there was no institutional linkage between project-designed organizations and the rural municipal councils, in PROGEDE-II, relations revolved around the need for the signature of the rural commune mayor for the recognition of those bodies. If rural municipal councils agreed to nominate some auditors to the association as requested by PROGEDE-II and the forest department, the latter would say there is an institutional relationship between the association and the rural municipal councils. Therefore, refusing to nominate auditors to the association was for rural municipal councils a way of declining any form of institutional linkage with the association since they disagreed with its establishment as it will be shown infra. So the short-term contracts rural commune mayors sign with associations to delegate them their authority over the forests are, in practice, the only form of institutional linkage between the two parties, under facilitation by PROGEDE-II and the forest department.

The same lack of accountability relation between forest management organizations at the municipal level and the commune is deeper between the departmental network of forest management organizations[8] and departmental councils – elected local governments at the administrative level of departments that compose an administrative region.

Discussion and conclusion

The chapter has examined how legal pluralism and institutional pluralism manifest, how they interact, and the implications for fragmentation. It has demonstrated that legal pluralism can be a twofold issue. Legal pluralism had usually been regarded as an external issue because of the coexistence of overlapping or merging different legal orders, mainly customary and state law (Benda-Beckmann, 1988; Griffiths, 1986; Merry, 1988). I have demonstrated that legal pluralism can be internal to a single legal order, such as state law as illustrated in this chapter. The internal character of legal pluralism manifests through the coexistence of multiple legal elements that are enforced (or invented) by the various state bodies from the national to the very local level, which act as separate domains of law within the state legal order per se.

Internal legal pluralism is not 'legal complexity,' which is the coexistence of laws of differing degrees of legitimacy – 'informal,' 'semi-informal' and 'formal' – in natural resource management (Simarmata, 2010: 117). Internal legal pluralism is affected by the autonomy of each of the legal (and regulatory) elements within a single legal order to serve as a law profitable to specific interests at a given time and space regardless of consistence of that legal order. Its emergence is enabled by the discretionary power each public administrative body is given to invent operational or technical regulations. The problem is that rather than seeking to put the law into practice, such regulations usually seek to create 'new rules of the game' or reorganize the game in a way that reasserts the power of the institutional body which holds regulatory powers while at the same time shrinking the other non-state institutional actors' authority over the resources. This chapter has shown how circulars and ministerial regulatory orders have been used by the forest agents at different levels of administration to take discretionary decision-making away from the hands of municipal council authorities on technical grounds.

Legal pluralism and institutional pluralism intersect concretely in social life. In the context of (external and internal) legal pluralism, the multiple institutional and organizational actors defined by decentralization reform and project interventions choose the legal resources that can support their present interests in the relevant sector. I have illustrated how classic legal pluralism has been used by the forest department to impose FMPs as autonomous legal instruments, although they were intended to be technical instruments only, despite the previous and long-standing state legislation on (forestry) lands.

When new norms are introduced to shift power positions, and authority over resources from actors that used to play key roles such as forest agents and village chiefs was withheld, the excluded former powerholders reinvent and revitalize any legal resource that may make them reemerge as legitimate players directly or indirectly. That was the case of village chiefs trying to control the choosing of the leaders of village committees and of rural commune mayors who are trying to control the anointment of leaders of the new associations introduced by PROGEDE-II. In this case, reinvention and revitalization occurred when the customary forest norms emerged as 'strategies for claiming and obtaining resources' (Meinzen-Dick and Pradhan, 2002: 2–3). Reinvention and revitalization have been described as 'an effort to transfer an important feature of the society or group considered to come from the realms of the past into present again' (Beyer, 2006: 148). Beyer (2006) contends that these conceptions of reinvention and revitalization hide the 'plurality of history.'

To advance local democracy in the sector of natural resources, the institutional pluralism that is exacerbated by the politics of recognition of development or environmental project interventions must obey the principle of institutional consistency. Institutional consistency is only possible when the newly created institutional bodies are put in a position to be responsive to the rightful institutional bodies that national legislation has granted the authority over the forests. In many countries, the rightful institutional bodies to run forest governance in decentralization contexts are elected local governments. Forest services must limit their intervention in decentralized forest governance only in giving technical advice and assistance to elected local governments for the implementation of decentralization in natural resources governance. For the time being, rather than providing technical advice and assistance to elected local governments, forest service agents are taking away the authority over the forests from elected local authorities through technical claims (Faye, 2015a, 2015b, 2018) weakly rooted in updated scientific results and modern technology.

The concept of institutional consistency may be a synonym to 'integral local development' which Ribot (2001b: 5) theorized as the merging of participatory bodies with the institutions resulting from or empowered by decentralization. It is only when institutional consistency guides institutional pluralism that we can avoid internal legal pluralism. While classical legal pluralism may introduce contradicting references among the coexisting social communities, internal legal pluralism is worse as it implies contradictions within the same social or administrative order depending on the changes in stakes and actors' interests about resources.

Even if institutional consistency is achieved, some mechanisms and procedures are needed to enable citizen engagement and participation in decision-making processes related to decentralized forest governments and especially as far as commercial resources are concerned. For that accountability to have a chance to occur, rural municipal councils may be put back in their lawful role of leading decisions that shape resources allocation and governance because a) they are regularly elected authorities and may want perform well to be reelected and b) they are the lawful institutional actors according to decentralization laws. This is said without ignoring the two everlasting weaknesses of the implementation of decentralization policy in the developing world: corruption in the form of bribing or privatization of public monies (Blundo, 2006) and absence of meaningful financial resources to take hold of the powers that are transferred including forest management (Faye, 2015a, 2015b, 2018; Ribot, 2002). These weaknesses may not suffice to reject decentralization policy as they are also present at the central government level; rather, scholars, practitioners and policy-makers should concentrate their efforts on joint initiatives to find pathways towards the resolving of these constraints. Although political decentralization is facing great challenges, local governments may be the most suited authorities to manage the forests with their territorial jurisdictions so as to avoid recurrent conflicts among actors (Faye, 2015) and contradictory regulations that divert all actors from the most important aim: resource sustainability! All these recommendations aim at framing the long work towards local democracy and well-being at the rural level.

Acknowledgments

The empirical data presented in this chapter derive from research supported mainly through a PhD studies grant offered to me by the Centre for Development and Environment of the University of Bern in Switzerland. Most recent fieldwork trips have been partially funded through a grant I earned in 2017–2018 from Open Society Foundations Fellowship Program in New York (Grant No. IN2017–35339). Thanks to Professor Jesse Ribot at American University of Washington and Nathalie Koenings, respectively, for commenting and editing early versions of this paper. I am also grateful to Professor Fiona Nunan at the University of Birmingham in the UK for inviting me to contribute this chapter to the book and for her constructive comments and criticism.

Notes

1 For legal anthropologists, law is any self-regulation of a 'semi-autonomous social field' among the various social fields that compose any society, not only state law (Griffiths, 1986: 38).
2 There is draft project law to add five more domains of power, among others are agriculture, pastoralism and animal production, and hydraulics.
3 Note that pluralism originated from a movement of reaction to the ancient Greeks' (Eleaticist) monism. It then spread out in social philosophy with the remarkable contribution of Aristotle, who pledged for a plurality of being and of types of knowledge (Johnstone and Marienthal-Maschler, 1962: 480). Thus, social pluralism was highly normative, seeking to overcome the disintegration of morality, and to justify a set of moral values for acceptable way of living together; morality becoming an integral part of faith.
4 However, in his latter works, Dahl seemed to reconsider his judgment about the propensity of 'polyarchy' for democracy, outlining four main inconveniences for pluralist systems: stabilization of injustices and inequalities, deformation of civic consciousness, distortion of the public agenda and misappropriation of the public functions (Dahl, 1982 cited in Manley, 1983: 371).

5 Similarly, Jacob (1997: 5) also conceived of 'multicentricity' or 'institutional piling up' (*empilement institutionnel ou multicentricité*) as being the plurality of decision-making loci, public and private, at the post-decentralization local level.
6 In case of loss of use rights for public purpose; those who could show *mise en valeur* could be paid for reparation.
7 An *arrondissement* is, after the village, the most rural level of the state territorial and administrative line. Neither belongs to the political decentralization line; rather, they belong to the administrative sector, led by appointed authorities, not elected as in the communes.
8 Associations of the communes within a department are regrouped into a departmental forest management association in each administrative department, and all those are also regrouped into regional forest management association in each region. Finally, all regional associations join each other to form the National Network of Forest Management Associations.

References

Benda-Beckmann, V. F. (1988) 'Comments on Merry', *Law & Society Review*, vol. 22, no. 5, pp. 897–902.

Benda-Beckmann, V. K. (1981) 'Forum shopping and shopping forums: Dispute processing in a Minangkabau village in West Sumatra', *Journal of Legal Pluralism*, vol. 19, pp. 117–159.

Beyer, J. (2006) 'Revitalization, invention and continued existence of the Kyrgyz *Aksakal* courts: Listening to pluralistic accounts of history', *Journal of Legal Pluralism and Unofficial Law*, vol. 38, no. 53–54, pp. 141–176.

Bierschenk, T. (1995) 'Rituels politiques et construction de l'identité ethnique des Peuls du Bénin', *Cahiers de sciences humaines*, vol. 31, no. 2, pp. 457–484.

Bierschenk, T. and Olivier de Sardan, J.-P. (2003) 'Powers in the village: Rural Benin between democratization and decentralization', *Africa*, vol. 73, no. 2, pp. 145–173

Blundo, G. (2006) 'Je n'ai pas besoin de ticket' Négociations des droits de marché et petite corruption dans les collectivités locales sénégalaises', in Claude Fay, Y. F. Koné and Cathérine Quiminal. Décentralsiation et pouvoirs en Afrique. En contrepoints, modèles territoriaux français. IRD Editions, Paris.

Bromley, D. W. (1991) *Environment and Economy. Property Rights and Public Policy.* Basil Blackwell, Cambridge, MA.

Dahl, R. A. and Lindblom, C. E. (1976) *Politics, Economics, and Welfare.* University of Chicago Press, Chicago.

Eckert, J. (2006) 'From subjects to citizens: Legalism from below and the homogenisation of the legal sphere', *Journal of Legal Pluralism and Unofficial Law*, vol. 38, no. 53–54, pp. 45–75.

Ensminger, J. (1992) 'A proper marriage: New institutional economic Anthropology', in J. Ensminger (ed) *Making a Market.* Cambridge University Press, Cambridge

Faye, P. (2006) 'Décentralisation, pluralisme institutionnelle et démocratie locale. Etude de cas de la gestion du massif forestier Missirah/Kothiary, Tambacounda (Sénégal)', *Séries de Monographies.* CODESRIA, Dakar.

Faye, P. (2015a) 'From recognition to derecognition: Hemming in democratic representation via technical claims. Responsive forest governance initiative', *Working Paper Series No. 17.* CODESRIA, Dakar.

Faye, P. (2015b) 'Choice and power: Resisting technical domination in Senegal's forest decentralization', *Forest Policy and Economics*, vol. 60, pp. 19–26.

Faye, P. (2018) 'Theorising derecognition of local government authorities as political injustice: The effects of technical claims in Senegal's forestry', *Conservation and Society*, vol. 15, no. 4, pp. 414–425.

Faye, P., Samb, A. C. D., Diop, D. and Ba, C. O. (2012) 'Enjeux fonciers à Ndiébène Gandiole. Les régulations à propos de la terre et du sel entre coutumes et modernités.' Rapport de diagnostic local, Programme "Negos-GRN." Initiative Prospective Agricole et Rurale (IPAR), Dakar, 65p.

Forsyth, T. (2009) 'Multilevel, multiactor governance in REDD+: Participation, integration and coordination', in A. Angelsen (ed.) *Realising Redd+: National Strategy and Policy Options*. CIFOR, Bogor.

Gellar, S. (1997) 'Conseils ruraux et gestion décentralisée des ressources naturelles au Sénégal, le défi: comment transformer ces concepts en réalité?', in P. Tersiguel et C. Becker (eds) *Le développement durable au Sahel*, 44–69.Karthala, Paris.

Griffiths, J. (1986) 'What is legal pluralism?', *Journal of Legal Pluralism and Unofficial Laws*, vol. 18, no. 24, pp. 1–55.

Haller, T. (2010) *Disputing the Floodplains. Institutional Change and the Politics of Resource Management in African Floodplains*. Brill, Leiden.

Idelman, E. (2008) *Le transfert des compétences en gestion des ressources naturelles aux communes rurales de la région de Kita (Mali). D'un encadrement étatique intégré à des logiques d'acteurs locaux*. Thèse de Doctorat en Géographie humaine, économique et régionale, Université Paris Ouest Nanterre- La Défense, Paris.

Jacob, J.-P. (1997) 'Etats et paysans face à la mise en oeuvre des politiques de décentralisation en Afrique', in J.-P. Jacob and G. Blundo (eds) *Socio-anthropologie de la décentralisation en milieu rural Africain*. Bibliographie sélective et commenté, IUED, Genève.

Johnstone, H. W. Jr. and Marienthal-Maschler, C. (1962) 'Skepticism and inferior knowledge: A note on Aristotle's pluralism', *Philosophy and Phenomenological Research*, vol. 22, no. 4, pp. 472–480.

Kanté, A. M. (2006) Décentralisation sans représentation. Le charbon de bois entre les collectivités locales et l'Etat. *Série de Monographies*. Programme: Pour une Gestion Décentralisée et Démocratique des Ressources Forestières au Sénégal, CIRAD-CODESRIA-WRI.

PROGEDE. (2004) Plan d'aménagement de la foret communautaire *Wulli/Laalo*.

Meinzen-Dick, R. S. and Pradhan, R. (2002) 'Legal pluralism and dynamic property rights', *CAPRi Working Paper 22*, International Food Policy Research Institute, Washington, DC.

Merry, E. M. (1988) 'Legal pluralism', *Law and Society Review*, vol. 22, no. 5, pp. 869–896.

Moore, K. M., Bertelsen, M. K., Diarra, L., Kodio, A., Cissé, S. and Wyeth, P. (2000) 'Natural resource management institution building in the decentralization context of West Africa: The SANREM CRSP approach', *Working Paper No. 01–02*. Paper for the Institutional Development Theme Session of the IFSA Symposium in Santiago, Chile.

Nader, L. (1972) *Up the Anthropologist: Perspectives Gained from Studying Up*. University of California Press, Berkeley.

Niane, D. T. (1989) *Histoire des Mandingues de l'Ouest*. Karthala-Arsan, Paris.

Nunan, F. (2018) 'Navigating multi-level natural resource governance: an analytical guide', *Natural Resources Forum*, vol. 42, no. 3, pp. 159–171.

Ostrom, E. (1999) *Self-governance and forest resources*. Centre for International Forestry Research (CIFOR), Indonesia.

Ribot, J. C. (2001a) *Historique de la gestion forestière en Afrique de l'Ouest. Ou: comment la science exclut les paysans*. World Resources Institute, Washington DC.

Ribot, J. C. (2001b) 'Integral local development: 'Accommodating multiple interests' through entrustment and accountable representation', *Journal of Agricultural Resources, Governance and Ecology*, vol. 1, no. 34, pp. 327–350.

Ribot, J. C. (2002) *Democratic Decentralization of Natural Resources: Institutionalizing Popular Participation*. World Resources Institute, Washington DC.

Ribot, J. C. and Faye, P. (2007) *Weex dunx and the Quota: The Plucking of Local Democracy in Senegal*. World Resources Institute.

Ribot, J. C., Chhatre, A. and Lankina, T. (2008) 'Institutional choice and recognition in the formation and consolidation of local democracy', *Conservation and Society*, vol. 6, no. 1, pp. 1–11.

Simarmata, R. (2010) 'Legal complexity in natural resource management in the frontier Mahakam Delta of Eastern Kalimatan, Indonesia', *Journal of Legal Pluralism and Unofficial Law*, vol. 62, pp. 115–146.

Wright, S. and Reinhold, S. (2011) "Studying through': A strategy for studying political transformation. Or sex, lies and British politics', in C. Shore, S. Wright and D. Però (eds) *Policy Worlds: Anthropology and the Analysis of Contemporary Power*. Berghahn Books, New York, Oxford.

11

IN SEARCH OF NEW MODES OF GOVERNANCE

The potential of conservation incentive payment policies to promote human–wildlife coexistence

Adam Pekor, Ingela Jansson, William Ole Seki, Dennis Rentsch, Göran Spong and Camilla Sandström

Introduction

The search for new paths to sustainable development is closely related to the search for new modes of governance. Most research in this area has focused on how to configure and improve governance processes through the inclusion of actors (Ansell and Gash, 2007) or how to design the 'rules of the game' that shape the interactions between actors (Decker et al., 2016; Mayntz, 2004; Ostrom, 2009; Rosenau, 1992). Although studies based on these two dimensions of governance—often denoted as the 'polity' and 'politics' dimensions—provide useful insights, they do not by themselves explain the full complexity of governance phenomena (Lange et al., 2013). What is often missing is the role of the policy dimension, that is, the specific content of governance, including both policy objectives and policy instruments (Orach and Schlüter, 2016; Treib et al., 2007).

Over the past two decades, a suite of innovative policy instruments known as conservation incentive payments (CIPs) have gained appeal as a way to promote human–wildlife coexistence by bridging the sharp disconnect that often exists between conservation policy and the sociocultural and economic interests of local people (Dickman et al., 2011; Kaczan et al., 2013; Marino and Pellegrino, 2018). For many communities, wildlife impose a significant financial burden, both directly, in the form of crop damage and livestock depredation, and indirectly, in the form of land-use restrictions designed to maintain suitable wildlife habitat (Langpap, 2006; Redpath et al., 2015). Yet the vast majority of the economic benefits derived from wildlife (e.g., through tourism and hunting concessions) are often captured by central governments or other outside actors (Nelson et al., 2010). As a result, it is often too costly for people to tolerate wildlife. CIPs help mitigate this problem by aligning the economic interests of local communities with conservation goals (Kaczan et al., 2013). (We use 'CIPs' as an umbrella term to include all policy instruments that promote conservation through the use of economic incentives, including payments for ecosystem services, performance payments, etc.) Under a CIP program, local stakeholders are paid directly for conserving species or their habitat, thereby minimizing (or eliminating) the financial burden imposed by wildlife (Dinerstein et al., 2013; Lapeyre, 2015). By transforming the economic impact of wildlife, CIPs can make conservation a financially viable option for local communities and improve outcomes for both people and wildlife.

Historically, the literature on governance, on the one hand, and on the characteristics of policy instruments like CIPs, on the other, have developed more or less independently. The two literatures should, however, be connected, since governance is best understood in the context of achieving policy goals and hence in relationship to policy instruments (Héritier, 2002). Moreover, governance systems are rarely capable of achieving policy goals without adequate policy instruments and actors to promote them. Accordingly, this chapter focuses on the nexus between governance and policy and demonstrates that each approach can help illuminate the other (Le Coq et al., 2015). In particular, we show that the introduction of a CIP program, a new kind of policy instrument, if successful, can facilitate a change from a top-down and hierarchical mode of governance to what could be described as a public–social partnership (Bjärstig and Sandström, 2017) or an interactive mode of governance (Lange et al., 2013).

To illustrate our argument, we use Tanzania's Ngorongoro Conservation Area (NCA) as a case study. In the NCA, intense human–lion conflict over the past several decades has threatened the government's stated objectives of conserving the NCA's natural resources, safeguarding the interests of its indigenous residents (all traditional pastoralists), and promoting tourism in the area (Jansson et al., forthcoming). In this regard, the NCA is not unique but rather representative of the problems faced by multiple-use commons around the world. We show how the introduction of a new policy instrument in the NCA has the potential to promote human–wildlife coexistence, while at the same time opening up the system for social inclusion and participation through interactive modes of governance.

Multidimensional approach to the analysis of governance

The implementation of 'new' environmental policy instruments like eco-taxes and payments for ecosystem services (PES) has led to both a 'fundamental transition' in environmental policy (Golub, 1998: xiii) and changes in modes of governance (e.g., from top-down to more interactive modes like public–private or public–social partnerships) (Lange et al., 2013). The main reasons for these changes have been a dissatisfaction with the effectiveness of traditional top-down regulatory instruments and the desire for greater local-level involvement through subsidiarity. These instruments are often 'proposed, designed and implemented by non-state actors, sometimes working alongside state actors, but sometimes also independently' (Jordan et al., 2005: 481).

As mentioned above, one such policy instrument is a CIP scheme, which takes the form of a voluntary transaction between at least one 'purchaser' (e.g., a government or nongovernmental organization) and one 'provider' (e.g., an individual or a community) in which the purchaser pays the provider if and only if the provider takes an agreed-on conservation action (e.g., refrains from hunting a target species) or achieves an agreed-on conservation goal (e.g., maintains a certain amount of the species' habitat) (Khalumba et al., 2014). The defining feature of a CIP scheme is that payment is conditioned on the performance of the conservation action and/or achievement of the conservation goal (Nelson et al., 2010). If the action is not taken and/or the goal is not achieved, no payment is made. Thus, a CIP scheme requires the establishment of a governing process whereby public and private actors, as interdependent partners, may benefit by combining their knowhow and expertise, but also financial and other resources, to deliver collective goods in a more efficient way, compared to top-down governance and privatization (Hodge and Greve, 2007).

A central theme in the discourse on CIPs has been economic efficiency (Ferraro and Kiss, 2002; Wunder, 2007). In particular, much attention has been focused on evaluating the costs of CIP schemes compared to other conservation approaches (Ferraro, 2001) and assessing how to achieve CIP efficiencies through the use of bidding processes and other market mechanisms (Groth, 2011). However, as a number of conservation practitioners have observed, determining the most efficient level of payment that will incentivize pro-conservation actions is just one issue to consider in implementing a CIP program, and perhaps not even the most important (Muradian et al., 2010). The successful design and implementation of a CIP scheme also requires coherence between policy aims and acceptance among involved actors, which are likely to be determined by factors relating to the three dimensions of governance: the local socio-economic, political and cultural conditions (politics); the level of collaboration, engagement, and fairness achieved in designing and implementing the scheme (polity); and, finally, the degree to which the implementing instrument is tailored to formal and informal institutions, laws, and norms (policy).

Politics: socioeconomic, cultural, and political context

As CIP schemes are, at bottom, interventions intended to conserve species or ecosystems, focusing on ecological considerations like the setting and monitoring of meaningful conservation targets is critical (Clifton, 2013; Jack et al., 2008; Prager et al., 2016; Sommerville et al., 2011). However, at the outset, equally important factors to consider are the socioeconomic, cultural, and political conditions in the target area, including current and previous modes of governance (Mishra et al., 2003). Indeed, without a basic understanding of the local people and the issues they face, program administrators are unlikely to be able to design a CIP scheme that the community will support. Despite the importance of understanding local stakeholders' needs and values, however, the vast majority of CIP programs pay little attention to them (Sorice and Donlan, 2015).

To gain a true understanding of local conditions, historical context is essential. For example, having dealt with inefficient, corrupt, or authoritarian governance systems may cause local stakeholders to be skeptical of participating in a CIP program or other conservation initiative (Bowden, 2004; Gonzalez and Jentoft, 2011; Petheram and Campbell, 2010). This was the case in Nicaragua's Pearl Cays, where local stakeholders were leery of establishing a marine protected area after years of having their fishing rights trampled on by what they viewed as a corrupt national government (Gonzalez and Jentoft, 2011). Likewise, in the South African province of KwaZulu-Natal, a history of top-down governance and fortress conservation made local stakeholders uneasy about establishing a community-based natural resource management (CBNRM) scheme (Bowden, 2004). Understanding and adapting the policy instrument to this type of historical political context can allow CIP programs to gain local support. For example, in southern Vietnam, Petheram and Campbell (2010) found that due to historical mistrust of government officials, local stakeholders were more comfortable entering into a CIP agreement, similar to a private social partnership, administered by an outside conservation organization.

While investigating and understanding local political conditions and culture takes time and resources, these efforts can yield lasting benefits for a conservation project. As Manjengwa and Murphree (2011) noted with respect to a CBNRM project in Zimbabwe, the time taken at the beginning of the project to understand the community with which it was being established was invaluable in building consensus and achieving successful outcomes later on. By contrast, a failure to account for context can not only doom a conservation project but even harm the local communities involved. For example, an attempt to establish a community-based conservancy (CBC) in northern Kenya without fully understanding the social, cultural, and political dynamics of the area exacerbated tensions between local ethnic groups over the distribution of CBC benefits and actually led to a 'full-scale guerrilla-type war' (Greiner, 2012).

Finally, independent of any CIP scheme, many local people undoubtedly have important relationships with and noneconomic reasons for valuing the environment. While setting up a CIP program, it is important to explore these motivations for conservation and ensure that the values are complemented, not supplanted, by incentive payments. Structuring a CIP program to align with, rather than 'crowd out,' noneconomic reasons for protecting wildlife may not only bolster the CIP program but also help ensure that the local conservation ethic persists even if the program ends (Baker et al., 2013; Muradian et al., 2010; Sorice and Donlan, 2015).

Polity: engagement, collaboration, and fairness

Recognizing local stakeholders as collaborators is also critical to the success of a community conservation endeavor like a CIP program (Ansell and Gash, 2007; Emerson et al., 2011; Engel et al., 2008; Mbaiwa and Stronza, 2011). Not only does working with local institutions and community members allow the program to benefit from their wealth of local knowledge and experience, but it is also essential to building trust, forging important relationships and establishing the legitimacy of the program (Gross-Camp et al., 2012).

In establishing a CIP program, local stakeholders should be engaged as early as possible in co-designing the various components of the program. 'If the process gets off on the wrong foot, it can be irreconcilable, preventing any meaningful engagement going forward' (Gerner et al., 2011). Thus, demonstrating to local people at the earliest phases that their views are valued and their concerns understood can help engender vital community support (Baker et al., 2013; Manjengwa and Murphree, 2011). Indeed, the incorporation of local stakeholders' preferences and recommendations is likely to be among the most important factors in determining the program's success (Gross-Camp et al., 2012). And, even if particular recommendations are not accepted, the process of engagement can itself generate goodwill with the community. Early engagement is also key to developing the specific components of the policy instrument, in terms of both governance and technical details. Here, it is important not only to agree on basic ecological targets but also develop what Emerson and Nabatchi (2015) define as a shared theory of change—the common objectives of the program and how they can be achieved.

Even if stakeholders are effectively engaged, incentive payments are likely to work only if they are fair and equitable to local people. Fairness relates both to the process of establishing a CIP program and to the outcome—the actual amount, form, timing and method of the payments made (Ansell and Gash, 2007; Gross-Camp et al., 2012). With regard to process, fairness means including local stakeholders in all aspects of decision making (García et al., 2014). Researchers on one PES project in Rwanda learned that involving local institutions and community members in designing the features of the program was essential to obtaining their buy-in. Just as important as the payments made under the program was the perception on the part of community members that they were being treated fairly throughout the process (Gross-Camp et al., 2012). Accordingly, taking steps to make sure that the poor, elderly, uneducated, and other marginalized groups are not excluded from the process by language and other socioeconomic barriers is critical (Clifton, 2013).

Fairness is also inherently tied to the equitability of how payments are distributed (Gross-Camp et al., 2012). In a number of PES and other community-conservation programs, the inequitable distribution of benefits in favor of local elites (and non-locals) was among the chief complaints of local stakeholders (Clifton, 2013; Kangalawe and Noe, 2012). Many CIP programs establish a method for distributing payments without explicitly considering the criteria used to assess equity (Pascual et al., 2010). However, because the interests and influence of program participants can vary widely, thinking critically about how equity will be determined is essential, both from an ethical perspective and in order to generate the greatest level of program support.

Policy: institutional fit and property rights

Finally, a CIP program is unlikely to be effective unless the policy is tailored to the local institutional and legal framework. This means, among other things, accounting for the strengths and weaknesses of local governance institutions—formal and informal—to determine the most effective means of establishing a CIP program. Functioning institutions are necessary to negotiate CIP terms, enter into an agreement, consolidate community support for the program, and distribute payments (Milne and Niesten, 2009). However, in many places where CIP programs have been established or are being considered, local governments and other civil institutions may not be functioning properly. In Cambodia, for example, one local committee that was responsible for administering a CIP agreement was nonfunctional for two years due to political infighting, which effectively suspended the program during that period (Milne and Niesten, 2009). In such cases, it may be appropriate for nongovernmental organizations (NGOs) and other non-state actors to help administer the CIP and assure local stakeholders that they will receive payments for providing the agreed-on services (Jack et al., 2008).

Similarly, because CIPs often condition payments on land use, understanding local property rights is also important. Conflicting claims to land may exacerbate conflict and lead to contested claims for payment (Greiner, 2012; Milne and Niesten, 2009). Thus, CIPs are more likely to be successful in places with well-defined and enforceable property rights. For example, in Tanzania, rural villages have a discrete corporate form, legal status, and geographic boundary (Nelson et al., 2010). Among other things, Tanzanian village councils are responsible for managing village lands and can own property, enter into contracts and take legal action to enforce their rights (Nelson et al., 2010). In northern Tanzania, these features of village governance have allowed for the successful implementation of a CIP program that pays villages for conserving a key wildlife dispersal area adjacent to Tarangire National Park (Nelson et al., 2010).

Case study: human–lion conflict in Tanzania's NCA

Overview of the NCA

The case we use to analyze the nexus between governance and policy is the NCA, an 8,300 km^2 multiuse protected area in northern Tanzania established in 1959 (Figure 11.1). Together with Serengeti National Park and surrounding areas, the NCA forms one of the largest savanna ecosystems in the world and has been designated as a UNESCO World Heritage Site and as part of the Serengeti-Ngorongoro Man and the Biosphere Reserve for its outstanding natural and cultural values. In addition to its high biodiversity, varied habitats, and abundant wildlife, the NCA is a particularly important site for lion conservation; among other things, the lions of the Ngorongoro Crater form one of the densest populations in Africa and have been studied extensively for over 50 years. The NCA's lions, along with its renowned natural features and archaeological sites,

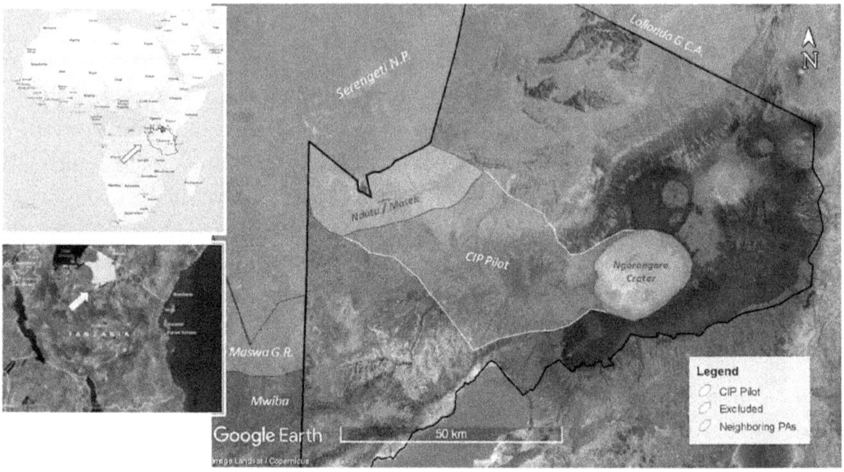

FIGURE 11.1 The Ngorongoro Conservation Area is a multiuse protected area and World Heritage Site located 180 km west of Arusha in northern Tanzania. The proposed area for piloting a conservation incentive payment program is situated between the Ngorongoro Crater in the east and the Ndutu/Masek area in the west.

make it one of Tanzania's premiere tourist destinations, drawing hundreds of thousands of tourists each year. However, the NCA is also home to approximately 100,000 people, the vast majority of whom are Maasai pastoralists, along with a smaller number of Datoga pastoralists and a few families of Hadzabe hunter-gatherers (NCA General Management Plan, 2006–2016).

Unique among Tanzania's protected areas, the NCA was formed for the joint purposes of conserving its natural resources, protecting the interests of the local people, and promoting tourism. Balancing these goals has proved challenging for the NCA Authority, the parastatal organization tasked with managing the area. While wildlife generate substantial revenue for the tourism industry, lions alone cause an estimated USD 70,000 to 75,000 in livestock losses per year, a significant burden on the NCA's residents, who rely largely on livestock for their livelihood and face rising levels of poverty. These losses often lead to retaliatory or preemptive killings, which (along with ritual killings) have caused lions to disappear from large parts of their historical range in the NCA. As a result, the lions of the Ngorongoro Crater have become increasingly isolated from the rest of the Serengeti ecosystem and have experienced a long history of close inbreeding (Kissui and Packer, 2004). In addition to the direct impact this conflict has had on local people and lions, it also has the potential to negatively impact the NCA's economy, since lions are an important driver of wildlife tourism in Africa (Willemen et al., 2015). Mitigating the NCA's human–lion conflict is thus essential for safeguarding the economic interests of the NCA's residents, conserving the local lion population, and ensuring that the NCA remains one of Africa's premiere tourist destinations.

Concern about the decline and fragmentation of the NCA's lion population, along with a recognition of the costs lions and other carnivores impose on the NCA's pastoralists, led one of us (IJ) to co-found KopeLion, a community-based NGO aimed at fostering long-term human–lion coexistence in the NCA. KopeLion utilizes the Lion Guardians model (Hazzah et al., 2014), pursuant to which resident pastoralist warriors are employed to mitigate human–wildlife conflicts in their home-communities and to monitor and protect lions. KopeLion has also continued and expanded on a long-term monitoring survey of lion demography in the NCA, in which each lion is recognized and followed throughout its lifetime. While we have observed some signs of increased tolerance towards lions and documented increased lion presence and movements across the NCA's landscape, there is a growing concern about the prospects of long-term coexistence due to the substantial costs lions impose on local people.

Despite the obvious importance of addressing the NCA's human–lion conflict, the situation has come to resemble a traditional collective action problem (Ostrom, 1990), where the lack of collaboration and coordination among key stakeholders has led to persistent conflict and where the present governance regime seems to lack appropriate instruments to promote conflict resolution (Jansson et al., forthcoming). Accordingly, stakeholders in the area are actively searching for new solutions and there is a window of opportunity to explore the possibility of implementing an alternative policy and governance arrangement. Beginning in 2017, we examined the drivers of the NCA's human–lion conflict and the feasibility of piloting a CIP program as a new policy and governance approach to mitigating it.

Investigating the relationship between governance and policy in the NCA's human–lion conflict

To analyze the potential for piloting a CIP program in the NCA, we reviewed secondary sources (e.g., legal and policy documents) and engaged local stakeholders in three stages. First, from June to September 2017, we conducted a feasibility study, in which we conducted interviews and focus groups (n = 29) with members of four key stakeholder groups: the NCA Authority; NCA residents; the Ngorongoro Pastoral Council (an elected advisory body that represents the NCA's residents); and tourism operators. We held meetings with NCA residents from six villages in KopeLion's study area—Kayepus, Oloirobi, Misigiyo, Longojoo, Endulen, and Nasporiong'—and, in each village, we engaged five different stakeholder groups: members of the village council; traditional Maasai leaders; village elders; current and recently retired Maasai warriors; and general community members (with a particular emphasis on the inclusion of women). With each stakeholder group, we obtained views on conservation and human–lion conflict in the NCA, opinions on the establishment of a lion CIP program, and preferences with respect to the design of such a program.

Second, following our feasibility study, in March 2018, we held a two-day joint-stakeholder workshop for the purpose of collaboratively outlining the framework for a potential pilot CIP program in the NCA. The workshop was attended by approximately 45 participants, including: 6 representatives from the NCA Authority; 4 representatives from the Ngorongoro Pastoral Council; 1 representative of the Ngorongoro District Council; 24 representatives from the villages engaged in our feasibility study; and 10 representatives from the tourism industry. The workshop was facilitated by several authors of this paper and a Maasai attorney. During the workshop, participants developed the framework for a potential lion CIP program in the NCA through a number of steps, including (1) identifying the most important issues to consider in establishing a CIP program, (2) developing various options for the design of a potential CIP program, and (3) conducting a SWOT (strength/weakness/opportunity/threat) analysis of the various options. Workshop participants interchangeably worked in small groups, designed to include participants from each stakeholder group, and presented their results in plenary sessions.

Third, based on the findings from our feasibility study and our workshop results, the NCA Authority conditionally approved the implementation of a pilot CIP program. Accordingly, we formed a CIP Design Committee consisting of representatives from all key stakeholder groups and held a second two-day workshop in April 2019 to collaboratively establish key terms and conditions of the pilot program.

On the design and development of a CIP program

Dissatisfaction with governance is a key driver of the NCA's human–lion conflict

The NCA's current conflict is embedded in a long history of land-use conflict, unclear property rights, mistrust, and a lack of benefit sharing (Jansson et al., forthcoming; Rogers, 2009). Indeed, our interviews and focus groups revealed that the negative attitudes towards lions shared by many local people are a product of dissatisfaction with the politics, polity, and policy aspects of governance in the NCA. First, many focus group participants explained that, historically, they have been excluded from the decision-making process. As one village council-person explained, 'Previously, decisions have been made without prior consultation, for example, making the Crater a no-go area. The NCA Authority's approach is top-down, not participatory. They issue dos and don'ts; even the police are better at communication.'[1] As a result, many people feel that wildlife have been prioritized over people. For example, one elder complained that

> people get killed and eaten by lions and there is no response, but if you kill a lion, there is a harsh response on you. The authorities have prioritized wildlife over us for a long time, which has hurt our views on wildlife.

Second, many participants expressed the view that, because the 'rules of the game' have limited their participation in the NCA's natural resource governance, they have not benefited sufficiently from the area's wildlife. As one elder stated, 'Ngorongoro is our home, our land; it produces a lot of money, but we have not benefited enough from it.' Another echoed, 'When I was young, the NCA Authority used to tell us "just don't kill [wildlife]." We didn't get any benefits though.' Or, as a warrior put it, 'Lions are on the [Tanzanian national] currency, which shows that the government sees them as a benefit. We should see a benefit too if we are to tolerate them.' Notably, participants were not focused specifically on current practices but rather on various actions taken since the NCA was formed in the 1950s: 'The NCA Authority's mandate includes considering the interests of the pastoralists, which they have not done for 60-plus years' (Elder).

Thus, as one interviewee summed up the conflict,

> Animals have no problems, it is people who have problems. It is between the community and the Authority, that is where the real conflict lies. Lion killing doesn't reflect conflict between people and lions as much as it reflects animosity against the NCA Authority, and dissatisfaction with the NCA Authority's approach to issues.

CIPs have the potential to positively impact governance in the NCA

Our research showed that a traditional, top-down compensation system, in which local people are partially or fully reimbursed for losses caused by carnivores, is not a viable model for mitigating human–lion conflict in the NCA due to problems with the verification of losses, transaction costs, fraud, and distrust between the NCA Authority and local residents. (In 2011, Tanzania adopted a policy of paying consolation (i.e., a small sum) for livestock losses caused by carnivores However, due to many of the complications noted above, that program was discontinued in the NCA.) Accordingly, stakeholders were nearly unanimous in the need for alternative policy options and open to the idea of CIPs. As one NCA Authority official explained,

> Compensation has failed. … We've been looking for a policy to help people for two to three years … There is clear viability [for a CIP program]; I see the NCA Authority supporting it. Incentive payments are a good way forward. … You don't fail until you stop trying.

Importantly, NCA Authority officials and community members alike recognized that not only would a CIP program allow the NCA's residents to profit from conserving lions, but it could also positively impact governance in the NCA in several ways. First, because a CIP program relies on (and rewards) people for their conservation efforts, it recognizes them as partners in conservation and important actors in the system. As one NCA Authority official explained, 'People should feel ownership of the wildlife—a CIP program can help do this.' Or, as another NCA Authority official stated, 'A CIP program is good because it provides measurable commitments on the part of the communities.'

In addition, many community members also noted that one of the most important benefits a CIP program could provide would be improved relations with the NCA Authority and an increased role in the decision-making process:

> [A CIP program] comes with the possibility of negotiations and discussions with the NCA Authority, which is itself a benefit because we've always dealt with these issues [without engagement from the NCA Authority]
>
> *(Elder)*

> I think [a CIP program] is the right path forward. The most beneficial thing apart from the money is stakeholders coming together for a unified purpose
>
> *(Warrior)*

NCA Authority officials also echoed this sentiment:

> Improved communication is a benefit that can be derived from a CIP program
>
> *(NCA Authority Official)*

> This type of a program could start to mend the relationship between the communities and the NCA Authority
>
> *(NCA Authority Official)*

Finally, a CIP program would also contribute to the empowerment of local communities since they would be able to make their own decisions about benefit distribution and other features of the program:

> A CIP program is preferable for two reasons: (1) benefits can come as long as people accept not to kill lions, and if we do well at mitigation, we can have cows and benefits; (2) through the dividend the community gets, we can choose to compensate individuals if we want.
>
> *(Community member)*

Developing a CIP policy in light of the NCA's socioeconomic, political, and cultural context

As explained above, in establishing a community conservation initiative like a CIP program, it is essential to understand the issues in context and from the perspective of the stakeholders involved. Here, it became clear from our focus groups and interviews that much of the hostility towards lions and other carnivores in Ngorongoro was really a proxy for hostility towards the NCA Authority. As one respondent put it, 'The NCA Authority are strangers to our land and they have antagonized us with wildlife; so any hate we have to wildlife is really because of the NCA Authority and not feeling unified with them' (Elder). Many respondents expressed that this sentiment has been building for decades, dating back to the removal of the Maasai from Serengeti National Park and the gazetting of the NCA in the 1950s: '[We're] concerned about what happened in Serengeti. White men came to conserve lions, then people got kicked out. This makes the pain fresh' (Village councilperson).

However, while many respondents expressed hostility towards the entity *officially* charged with conserving the NCA's wildlife, those feelings exist within the context of a strong conservation ethic shared by many of the NCA's residents. Indeed, despite feeling like they have benefited little from wildlife, many respondents expressed that they have conserved wildlife because of Maasai traditions and customs:

> Our people were the original conservationists. We lived here before it was gazetted. We conserve, we don't kill wildlife
>
> *(Elder)*

> [When there was a] rhino that came out of the Crater, the people herded it, called the NCA Authority, guarded it; they didn't try to kill it and take its horn. It showed that we are conservationists
>
> *(Elder)*

As a result, many respondents acknowledged the inherent value of human–wildlife coexistence in the NCA and the need for new solutions:

> It's not about whether we like [wildlife] or not, it's about everything deserving a place.
>
> *(Traditional leader)*

> Ngorongoro's uniqueness is because of people and wildlife. Our ancestors conserved, and we carry on with it despite the burden
>
> *(Community member)*

> We should not focus on *whether* Ngorongoro should be a mixed-use area, but focus on *ways* for coexistence. The eyes of the world are on us
>
> *(Traditional leader)*

Accordingly, our research showed that many local people have important relationships with and noneconomic reasons for valuing the environment, which should be taken into consideration in designing a CIP program.

Collaboratively designing a fair and equitable CIP policy

Although there was widespread enthusiasm for the potential benefits of introducing a CIP program in the NCA, participants raised a number of concerns about ensuring the fairness and equitability of the policy. In particular, these concerns related primarily to (1) the conditions for earning benefits and (2) the manner in which benefits would be distributed and managed. However, each of these issues were collaboratively addressed by stakeholders, who developed agreed-on solutions during the joint-stakeholder CIP Design Committee workshop held in April 2019.

Conditions for earning benefits

Under the proposed CIP program, communities will earn benefits based on the presence of lions on ward land, as determined by KopeLion's continued lion monitoring activities, including the visual recognition of individual lions, the analysis of lion spoors (e.g., tracks, scat, hair), and the use of GPS collar data, lion call-ups, and motion-triggered camera traps. Benefit levels will be calculated based on the number of unique, individually recognized lions on ward land each month, with a substantial penalty for any lion killings (i.e., forfeiture of benefits for a certain number of months).

Community concerns with this proposed framework focused primarily on the possibility that they might not be credited for all lions observed on ward land each month. For example, with respect to the use of camera traps, one community member asked, 'How will the community know what the cameras are showing if only KopeLion gets to look at them?' A warrior put it more bluntly: 'How will we know we're not being cheated based on the number of lions?' In response, it was suggested by other respondents that each village could designate representatives to participate in the photo review process:

> As long as it's our person who reviews the pictures then we'll trust it. But if it was someone from far away, we wouldn't trust it necessarily
>
> *(Community member)*

> I would trust the process if we were involved in looking at the cameras
>
> *(Village councilperson)*

In addition, focus group participants generally agreed that, if they were involved in the verification process and no lions were observed on ward land during a given month, they would accept as fair that they would not receive a benefit for that month:

> [It's like] a cow, [which] usually gives birth once a year. Some years it doesn't, but you know you have the chance to get a calf the next year
>
> *(Community member)*

> This is completely reasonable; money shouldn't come for free. Cameras will show that good care of lions is being taken. A hard worker is always rewarded by his work
>
> *(Community member)*

Accordingly, at the CIP Design Committee workshop held in April 2019, it was agreed that each village participating in the CIP program would designate a certain number of CIP Community Liaisons whose responsibilities would include participating in the lion verification process. Not only did this solution address concerns about fairness, but the inclusion of community members in these biotic surveys and other aspects of program management also has the potential to reinforce support for the program (Emerson and Nabatchi, 2015; García et al., 2014; Gonzalez and Jentoft, 2011).

In addition to concerns with the benefit verification process, there was initial disagreement among community members about the proposed penalties for lion killings under the program. On the one hand, most respondents recognized that earning benefits under the proposed program entailed a commitment not to kill lions. As one community member noted, 'It is fair to require no lion killings [in exchange for benefits].' Indeed, several participants noted that if benefits under a CIP program were conditioned on a commitment not to kill lions, social pressure and cultural norms would limit such killings:

> The rest of the community would view a person badly if they defied the common purpose
>
> *(Warrior)*

> The advantage of incentives is that the community would not protect the person who killed a lion. It would also help people discourage hunts because the community will put pressure on those who want to hunt
>
> *(Villager)*

However, other community members raised concerns about the fairness of such a categorical prohibition. Specifically, a number of respondents emphasized the dangers that lions pose to people and livestock and urged that people should be permitted to defend themselves and their livestock without being penalized:

> To accept that there is no lion killing at all would mean that I couldn't kill a lion while I was right there to defend my livestock
>
> *(Elder)*

> If a lion attacks you, you have to defend yourself
>
> *(Warrior)*

Accordingly, a compromise was reached and approved by the CIP Design Committee whereby no penalty would be imposed for lion killings in cases of self-defense, that is, where the persons involved reasonably believed that the lion posed an immediate threat to the safety of any person, livestock, or other domestic animal. It was further agreed that, for purposes of the CIP program, the determination of whether any killing was in self-defense would be made by a majority determination of one representative of the NCA Authority, one representative of the community, and one representative of KopeLion.

Distributions of benefits

Discussions on how the benefits earned under the CIP program should be distributed and managed also focused on issues of fairness and equity. While people in positions of authority (village council members, traditional leaders) often suggested that funds earned under a CIP program should be managed by village leadership, others expressed concerns about corruption and transparency and suggested that benefits be given in the form of services or the direct funding of projects, so that CIP funds are not managed by local leaders:

> I would prefer that money be given in the form of a service, not money to the government because of corruption
>
> *(Community member)*

> We've experienced that money received never gets used for the intended purpose, so the receipt of services is better
>
> *(Elder)*

The question of individual versus communal benefits was also a central issue. While it was assumed that individual benefits would have the potential to reach most or all members of a community, it was recognized that they might be too insignificant to meaningfully impact people's lives. By contrast, communal benefits could be more impactful but might not be used or valued by all members of the community. Given these trade-offs, community members were split on the manner in which benefits should be provided. Those who favored individual benefits emphasized the individual nature of losses caused by lions. As one community member pointed out, 'a CIP program is a community thing, but what if I lose a cow and it was my only cow? How would a CIP program help me?'

Still, many others favored communal benefits that could bring meaningful projects to their village, especially since the benefits of such projects could be long-lasting:

> I would see it long-term. I would feel the pain [of livestock losses], but would recognize the long-term value of the benefits, and would support the program also because it is a community decision. Long-term benefits would outweigh short-term pain of losses
>
> *(Village councilperson)*

At the April 2019 workshop, the CIP Design Committee considered how to balance the interests of community self-determination with respect to the use of benefits, on the one hand, versus ensuring that benefits earned under the CIP pilot program are used in a manner that is both appropriate and impactful, on the other. Ultimately, a consensus was reached that (a) the memorandum of understanding (MOU) establishing the CIP pilot program should include a general framework for the permissible uses of funds earned under the program; (b) funds should be disbursed directly to participating communities to be used in any manner consistent with the MOU; (c) participating communities should be required to account for all funds received under the program; and (d) a penalty (i.e., the forfeiture of future benefits) should be imposed if any funds are used in a manner inconsistent with the MOU.

In addition to these formal reporting requirements (e.g., the submission of budgets and receipts), recommendations for ensuring fairness and equity in the use of funds included requiring communities to publicly display how CIP funds were used; requiring communities to report on the use of CIP funds at public meetings; and, utilizing CIP Community Liaisons to survey and report back on community satisfaction with the use of CIP funds in their villages. If final approval to pilot the CIP program is granted, how participating communities resolve the question of individual versus communal benefits and whether communal benefits are sufficient to promote tolerance of lions in the face of individual livestock losses will be key issues to evaluate.

Institutional fit and funding of a CIP program

Designing a CIP policy for the NCA required recognition of the constraints imposed by funding limitations, administrative capacity and national law and policy. With respect to funding, NCA Authority officials expressed mixed views on the willingness of the NCA Authority Board of Directors to contribute funding for a CIP program. As one official noted, 'Money will be a hurdle, getting the NCA Authority on board. Money already goes to the PC [Pastoral Council], schools and other activities ... Asking for more money will be difficult.' However, that the same official also suggested that the NCA Authority might be willing to relocate its budget for the suspended consolation program to support a CIP pilot instead. Indeed, in follow-up communications since our initial feasibility study, NCA Authority officials have increasingly asserted that, because it is the NCA Authority's responsibility to address the area's human–lion conflict, if a CIP program is to be piloted, it is incumbent on the NCA Authority to support the project both financially and administratively.

With respect to program administration, however, all stakeholders recognized that the communities' long history of distrust of the NCA Authority would limit faith in the program if it were managed by the agency. Likewise, many stakeholders recognized that the NCA Authority was not prepared to handle certain administrative activities, such as lion monitoring and verification, fundraising, external communications and dispute mediation. Accordingly, all stakeholders agreed that it would be necessary for some independent NGO to be involved in administering the program. However, as noted above, NCA Authority officials increasingly asserted the importance of the agency's participation in program administration. Likewise, stakeholders began to recognize the importance of communities' participation, not just by providing input in the program's design, but also in administering it as well. Accordingly, during the April 2019 workshop, a consensus was reached that KopeLion should manage the day-to-day operations of the program and that the joint-stakeholder CIP Design Committee should serve as a permanent committee, to meet twice per year for the purpose of adaptively managing the program to address new and unforeseen challenges.

One final issue that was addressed concerning institutional fit was how to establish a CIP program within Tanzania's existing legal and policy framework. Because Tanzania's national policy is to pay consolation for livestock losses caused by wildlife, some NCA Authority officials expressed concern that the NCA Authority Board and/or the Ministry of Natural Resources and Tourism might oppose the establishment of an alternative program in the NCA:

> The Tanzanian government may not wish to see a double standard between the NCA and other areas
>
> *(NCA Authority Official)*

> The challenge will come from the parent ministry which may feel like it has its own policies already. It may not want the NCA Authority to come up with its own policies
>
> *(NCA Authority Official)*

However, through a review of the relevant statutes and regulations, it was determined that the establishment of a CIP program in the NCA would be consistent with Tanzanian law and policy. While Tanzania's Wildlife Conservation Act, 2009, provides that the government may implement a consolation program, that statute does not establish consolation payments as the only permissible means of promoting human–wildlife coexistence in Tanzania. Indeed, piloting a CIP program in the NCA would directly promote the national policy objectives to 'protect and conserve wildlife resources,' 'encourage, promote and facilitate active involvement and participation of local and traditional communities in the sustainable management, use and conservation of wildlife,' 'facilitate greater public awareness of the cultural, economic and social benefits for conserving wildlife,' and 'mitigate human-wildlife conflicts wherever they occur' Wildlife Conservation Act, 2009 § 5-(1)(a), (g), (j), and (k).

In fact, several officials noted that because of the NCA's unique status as the only Tanzanian protected area in which people live, the NCA Authority Board and the Ministry would be open to piloting a CIP program there.

> There are many national policies and political considerations … but the NCA is unique. This may be a reason to treat the NCA differently than other places. Any proposal would have to spell out clearly that it is a pilot and that the NCA is unique …. The NCA is seen as a model for other protected areas to try new things …
>
> *(NCA Authority Official)*

By the conclusion of the April 2019 workshop, it was generally agreed by the NCA Authority that its unique legal status provided the flexibility necessary to pilot a new policy instrument like a CIP program.

Insights from the policy literature for natural resource governance

The nexus between the three dimensions of governance, here illustrated in the context of a particular policy instrument—a CIP program—and its application to the case of the NCA, provides us with important lessons for the analysis of natural resource governance. First, the increasing complexity and dependencies between actors requires more coordination and collaboration to address complex governance problems, such as human–wildlife conflicts. This has, in turn, opened a window for the design of new policies and policy instruments that are dependent on participatory or inclusive modes of governance (Howlett and Mukherjee, 2017; Howlett and Rayner, 2007). Our case study shows that the co-design of a CIP program, which takes into account both historical and current political and sociocultural conditions, provides an opportunity to move from what can be characterized as a political deadlock, in our particular case, towards a more sustainable situation. However, to achieve long-term success, it is necessary to build capacity, that is, the ability of individuals and institutions to effectively make and implement decisions and perform various functions. Capacity building is needed at the individual level, to prepare the actors involved to meaningfully contribute to the development of the program; at the organizational level, to allow governmental and nongovernmental organizations to manage key resources, including leadership, information, and funding; and, finally, at the societal level, to empower local communities to support the program's development.

Second, although theories of public policy and theories of governance both focus on the relationships between policy making and its consequences, the links between the three dimensions of governance are rarely made explicit. Often, how these relationships impact on outputs and outcomes are not taken into account. Hence, both theoretical frameworks would benefit from adopting a multidimensional approach, giving consideration to all three dimensions of governance, that is, politics, polity, and policy. From an empirical perspective, a multidimensional perspective is equally important, since a pilot program like the one examined in our chapter needs to be continuously evaluated, not only to meet its overarching objectives but also to provide the information necessary for adaptive governance.

Finally, a CIP program is unlikely to be a 'silver bullet' solution to any particular human–wildlife conflict. While CIPs can help align the interests of local people with conservation goals, it is necessary to carefully consider the relationship between local people and conservation and the potentially negative impacts of external national and global conservation interests on local communities' rights and livelihoods. Optimally, consideration should be given to implementing CIP programs together with other measures that have the potential to strengthen and complement each other and improve both conservation and development goals.

Note

1 To promote a full and frank discussion during our interviews and focus groups, we advised all participants that we would not publish any personally identifiable information. Accordingly, we attribute quotes here based only on the respondent's stakeholder group ('traditional leader,' 'elder,' 'NCA Authority official,' etc.).

References

Ansell, C. and Gash, A. (2007) 'Collaborative governance in theory and practice', *Journal of Public Administration Research and Theory*, vol. 18, no. 4, pp. 543–571.

Baker, J., Bitariho, R., Gordon-Maclean, A., Kasoma, P., Roe, D., Sheil, D., Twinamatsiko, M., Tumushabe, G., van Heist, M. and Weiland, M. (2013) 'Linking protected area conservation with poverty alleviation in Uganda: Integrated conservation and development at Bwindi impenetrable National Park', in Smith, J. (ed.) *National parks: Sustainable development, conservation strategies, and environmental impacts*. Nova Publishers, New York, NY.

Bjärstig, T. and Sandström, C. (2017) 'Public-private partnerships in a Swedish rural context-A policy tool for the authorities to achieve sustainable rural development?', *Journal of Rural Studies*, vol. 49, pp. 58–68.

Bowden, A. P. D. (2004) 'Identifying criteria for the successful implementation of community-based conservation initiatives: Evidence from two case studies in KwaZulu-Natal, South Africa', Master's thesis, University of KwaZulu-Natal, South Africa.

Clifton, J. (2013) 'Compensation, conservation and communities: An analysis of direct payments initiatives within an Indonesian marine protected area', *Environmental Conservation*, vol. 40, no. 3, pp. 287–295.

Le Coq, J.-F., Froger, G., Pesche, D., Legrand, T. and Saenz, F. (2015) 'Understanding the governance of the payment for environmental services programme in Costa Rica: A policy process perspective', *Ecosystem Services*, vol. 16, pp. 253–265.

Decker, D., Smith, C., Forstchen, A., Hare, D., Pomeranz, E., Doyle-Capitman, C., Schuler, K. and Organ, J. (2016) 'Governance principles for wildlife conservation in the 21st century', *Conservation Letters*, vol. 9, pp. 290–295.

Dickman, A. J., Macdonald, E. A. and Macdonald, D. W. (2011) 'A review of financial instruments to pay for predator conservation and encourage human–carnivore coexistence', *Proceedings of the National Academy of Sciences*, vol. 108, no. 34, pp. 13937–13944.

Dinerstein, E., Varma, K., Wikramanayake, E., Powell, G., Lumpkin, S., Naidoo, R., Korchinsky, M., Del Valle, C., Lohani, S., Seidensticker, J., Joldersma, D., Lovejoy, T. and Kushlin, A. (2013) 'Enhancing conservation, ecosystem services, and local livelihoods through a wildlife premium mechanism', *Conservation Biology*, vol. 27, no. 1, pp. 14–23.

Emerson, K. and Nabatchi, T. (2015) 'Evaluating the productivity of collaborative governance regimes: A performance matrix', *Public Performance & Management Review*, vol. 38, pp. 717–747.

Emerson, K., Nabatchi, T. and Balogh, S. (2011) 'An integrative framework for collaborative governance', *Journal of Public Administration Research and Theory*, vol. 22, no. 1, pp. 1–29.

Engel, S., Pagiola, S. and Wunder, S. (2008) 'Designing payments for environmental services in theory and practice: An overview of the issues', *Ecological Economics*, vol. 65, pp. 663–674.

Ferraro, P. J. (2001) 'Global habitat protection: Limitations of development interventions and a role for conservation performance payments', *Conservation Biology*, vol. 15, no. 4, pp. 990–1000.

Ferraro, P. J. and Kiss, A. (2002) 'Direct payments to conserve biodiversity', *Science*, vol. 298, pp. 1718–1719.

García, C., Tavera-Escobar, H., Vieira, C., Rincón, C. and Rentería, E. (2014) 'Fostering ethno-territorial autonomy: A Colombian case study of community-based conservation of Mangroves', *Journal of Latin American Geography*, vol. 13, no. 2, pp. 117–152.

Gerner, J., Heurich, M., Günther, S. and Schraml, U. (2011). 'Red deer at a crossroads—An analysis of communication strategies concerning wildlife management in the 'Bayerischer Wald' National Park', Germany', *Journal for Nature Conservation*, vol. 19, no. 5, pp. 319–326.

Golub, J. (1998) *New Instruments for Environmental Policy in the EU*. Routledge, London.

Gonzalez, C. and Jentoft, S. (2011) 'MPA in labor: Securing the pearl cays of Nicaragua', *Environmental Management*, vol. 47, no. 4, pp. 617–629.

Greiner, C. (2012) 'Unexpected consequences: Wildlife conservation and territorial conflict in Northern Kenya', *Human Ecology*, vol. 40, no. 3, pp. 415–425.

Gross-Camp, N., Martin, A., McGuire, S., Kebede, B. and Munyarujaza, J. (2012) 'Payments for ecosystem services in an African protected area: Exploring issues of legitimacy, fairness, equity and effectiveness', *Oryx*, vol. 46, no. 1, pp. 24–33.

Groth, M. 2011. 'Cost-effective biodiversity conservation: Procurement auctions and payment-by-results', *Eurochoices*, vol. 10, no. 2, pp. 32–37.

Hazzah, L., Dolrenry, S., Naughton, L., Edwards T. T. C., Mwebi, O., Kearney, F. and Frank, L. (2014) "Efficacy of two lion conservation programs in Maasailand, Kenya", *Conservation Biology*, vol. 28, no. 3, pp. 851–860.

Héritier, A. (2002) New modes of governance in Europe: Policy making without legislating? *IHS Political Science Series: 2002, No. 81.* [Working Paper].

Hodge, G. A. and Greve, C. (2007) 'Public–private partnerships: An international performance review', *Public Administration Review*, vol. 67, no. 3, pp. 545–558.

Howlett, M. and Mukherjee, I. (2017) 'Policy design: From tools to patches', *Canadian Public Administration*, vol. 60, no. 1, pp. 140–144.

Howlett, M. and Rayner, J. (2007) 'Design principles for policy mixes: Cohesion and coherence in 'new governance arrangements', *Policy and Society*, vol. 26, no. 4, pp. 1–18.

Jack, B., Kousky, C. and Sims, K. (2008) 'Designing payments for ecosystem services: Lessons from previous experience with incentive-based mechanisms,' *Proceedings of the National Academy of Sciences*, vol. 105, no. 28, pp. 9465–9470.

Jansson, I., Pekor, A., Spong, G. and Sandstrom, C. (forthcoming) 'Mapping co-existence in crowded landscapes – Lions, pastoralism and tourism in Ngorongoro Conservation Area, Tanzania'.

Jordan, A., Wurzel, R. K. W. and Zito, A. (2005) 'The rise of 'new' policy instruments in comparative perspective: Has governance eclipsed government?', *Political Studies*, vol. 53, pp. 477–496.

Kaczan, D., Swallow, B. and Adamowicz, W. (2013) 'Designing payments for ecosystem services (PES) program to reduce deforestation in Tanzania: An assessment of payment approaches', *Ecological Economics*, vol. 95, pp. 20–30.

Kangalawe, R. Y. M. and Noe, C. (2012) 'Biodiversity conservation and poverty alleviation in Namtumbo District, Tanzania', *Agriculture, Ecosystems and Environment*, vol. 162, pp. 90–100.

Khalumba, M., Wünscher, T., Wunder, S., Büdenbender M. and Holm-Müller, K. (2014) 'Combining auctions and performance-based payments in a forest enrichment field trial in Western Kenya', *Conservation Biology*, vol. 28, pp. 861–866.

Kissui, B. M. and Packer, C. (2004) 'Top-down population regulation of a top predator: Lions in the Ngorongoro Crater', *Proceedings of the Royal Society B: Biological Sciences*, vol. B271, pp. 1867–1874.

Lange, P., Driessen, P. P. J., Sauer, A., Bornemann, B. and Burger, P. (2013) 'Governing towards sustainability—Conceptualizing modes of governance', *Journal of Environmental Policy and Planning*, vol. 15, no. 3, pp. 403–425.

Langpap, C. (2006) 'Conservation of endangered species: Can incentives work for private landowners?', *Ecological Economics*, vol. 57, no. 4, pp. 558–572.

Lapeyre, R. (2015) 'Wildlife conservation without financial viability? The potenYtial for payments for dispersal areas' services in Namibia', *Animal Conservation*, vol. 18, no. 1, pp. 14–15.

Manjengwa, J. and Murphree, M. (2011) 'Local level scenario planning, iterative assessment and adaptive management', Final Technical Report. Center for Applied Social Sciences, University of Zimbabwe.

Marino, D. and Pellegrino, D. (2018) 'Can payments for ecosystem services improve the management of Natura 2000 Sites? A contribution to explore their role in Italy', *Sustainability*, vol. 10, no. 3, p. 665.

Mayntz, R. (2004) 'Governance im modernen Staat', in Benz, A. (ed.) *Governance—Regieren in komplexen Regelsystemen. Governance*. VS Verlag für Sozialwissenschaften.

Mbaiwa, J. E. and Stronza, A. L. (2011) 'Changes in resident attitudes towards tourism development and conservation in the Okavango Delta, Botswana', *Journal of Environmental Management*, vol. 92, no. 8, pp. 1950–1959.

Milne, S. and Niesten, E. (2009) 'Direct payments for biodiversity conservation in developing countries: Practical insights for design and implementation', *Oryx*, vol. 43, pp. 530–540.

Mishra, C., Allen, P., McCarthy, T. O. M., Madhusudan, M. D., Bayarjargal, A. and Prins, H. H. T. (2003) 'The role of incentive programs in conserving the snow leopard', *Conservation Biology*, vol. 17, no. 6, pp. 1512–1520.

Muradian, R., Corbera, E., Pascual, U., Kosoy, N. and May, P. H. (2010) 'Reconciling theory and practice: An alternative conceptual framework for understanding payments for environmental services', *Ecological Economics*, vol. 69, no. 6, pp. 1202–1208.

Nelson, F., Foley, C., Foley, L. S., Leposo, A., Loure, E., Peterson, D., Peterson, M., Peterson, T., Sachedina, H. and Williams, A. (2010) 'Payments for ecosystem services as a framework for community-based conservation in Northern Tanzania', *Conservation Biology*, vol. 24, no. 1, pp. 78–85.

Orach, K. and Schlüter, M. (2016) 'Uncovering the political dimension of social-ecological systems: Contributions from policy process frameworks', *Global Environmental Change*, vol. 40, pp. 13–25.

Ostrom, E. (1990) *Governing the commons: The evolution of institutions for collective action*. Cambridge University Press, Cambridge.

Pascual, U., Muradian, R., Rodríguez, L. C. and Duraiappah, A. (2010) 'Exploring the links between equity and efficiency in payments for environmental services: A conceptual approach', *Ecological Economics*, vol. 69, no. 6, pp. 1237–1244.

Petheram, L. and Campbell, B. (2010) 'Listening to locals on payments for environmental services', *Journal of Environmental Management*, vol. 91, no. 5, pp. 1139–1149.

Prager, C., Varga, A., Olmsted, P., Ingram, J., Cattau, M., Freund, C., Wynn-Grant, R. and Naeem, S. (2016) 'An assessment of adherence to basic ecological principles by payments for ecosystem service projects', *Conservation Biology*, vol. 30, no. 4, pp. 836–845.

Redpath, S., Gutiérrez, R., Wood, K. and Young, J. (Eds.) (2015) *Conflicts in Conservation: Navigating Towards Solutions (Ecological Reviews)*. Cambridge University Press, Cambridge.

Rogers, P. J. (2009) 'International conservation governance and the early history of the Ngorongoro conservation area, Tanzania', *Global Environment*, vol. 4, pp. 78–117.

Rosenau, J. N. (1992) 'Normative challenges in a turbulent world', *Ethics and International Affairs*, vol. 6, no. 1, pp. 1–19.

Sommerville, M., Milner-Gulland, E. and Jones, J. (2011) 'The challenge of monitoring biodiversity in payment for environmental service interventions', Biological Conservation, vol. 144, no. 12, pp. 2832–2841.

Sorice, M. and Donlan, J. (2015) 'A human-centered framework for innovation in conservation incentive programs', *Ambio*, vol. 44, pp. 788–792.

Treib, O., Bähr, H., and Falkner, G. (2007) 'Modes of governance: Towards a conceptual clarification', *Journal of European Public Policy*, vol. 14, no. 1, pp. 1–20.

Willemen, L., Cottam, A. J., Drakou, E. G. and Burgess, N. D. (2015) 'Using social media to measure the contribution of red list species to the nature-based tourism potential of African protected areas', *PLoS ONE*, vol. 10, no. 6, e0129785.

Wunder, S. (2007) 'The efficiency of payments for environmental services in tropical conservation', *Conservation Biology*, vol. 21, no. 1, pp. 48–58.

12

CONCLUSION

Working towards diverse, flexible and inclusive natural resource governance

Fiona Nunan

Introduction

The preceding chapters introduced a range of approaches that have been, or could be, used to analyse the governance of renewable natural resources and illustrated the application of these through case studies. This chapter compares and contrasts the approaches to identify what it is about governance that has been analysed and what common themes emerge from the analysis.

Table 12.1 presents a summary of Chapters 2 to 11, in terms of the approach taken, the focus of the analysis, methods used to generate and analyse data and key insights arising from the reflection on an area of theory, a framework or approach and their application. Each of these aspects is considered in turn in this chapter, with a conclusion that brings together key themes and findings from across the chapters.

Frameworks, theories and approaches

The frameworks, theories and approaches introduced and applied in Chapters 2 to 11 include the following: political economy of institutions; social network analysis; the social-ecological systems framework, grounded in collective action theory; critical institutionalism and institutional bricolage; mainstream institutionalism (or common property theory) and critical institutionalism; polycentricity; political ecology; literature on the anthropologies of the African state, public administration studies, organisation studies, development studies and anthropology; legal and institutional pluralism; and the interface between politics, polity and policy.

In Chapter 2, Kashwan draws on a political economy of institutions (PEI) perspective, which 'takes a critical understanding of institutions and governance

TABLE 12.1 Comparison of approaches and insights to analysing governance

Chapter/Approach	*Focus of analysis*	*Methodologies*	*Insights*
2/Political economy of institutions (PEI) approach	Governance indicators have not asked fundamental questions such as governance for what, by whom and under what conditions	Analysis of documents associated with three governance assessment frameworks	Develops and applies a meta-framework with five dimensions: 1. Specification of goals 2. Representation of stakeholders 3. Monitoring and enforcement 4. Credible commitment of state power 5. Powers and counter-powers 'Governance is essentially a 'political' matter'
3/Social network analysis (SNA)	Analysing participation in governance through analysis of relations within networks, asking whether participatory forums achieve their objectives of inclusion and facilitating horizontal deliberation	Questionnaires completed by participants in a water basin council, with modelling of the networks using Exponential Random Graph Models	• Social dynamics such as reciprocity and popularity dominate collaboration in the participatory forum. • SNA can be used to identify who is influential and which actors are marginalized and why, and with what implications for natural resource management.
4/Social-ecological systems framework (SESF)	Uses the SESF as a diagnostic tool to examine variables that influence collective action and cooperation within natural resource governance	Qualitative interviews undertaken, with coding of responses guided by the SESF	• Many variables were identified as influencing governance • Policy solutions are difficult to simplify given the number of influential variables and their interactions • Capacity building may be needed to regain trust and social energy, particularly through improved communication • The SESF cannot be considered as 'final', critiques and limitations are noted

(*Continued*)

Chapter/Approach	*Focus of analysis*	*Methodologies*	*Insights*
5/Critical institutionalism and institutional bricolage	How socially-embedded and bureaucratic institutions interact and shape the practice and outcomes of governance	Qualitative interviews with a range of stakeholders	• Socially-embedded institutions affect participation of actors in co-management through power relations, social norms and social and economic relations. • Bricolage affects the practice and outcomes of fisheries co-management, largely by enabling more powerful actors to remain dominant at the local level and enabling illegalities to continue.
6/Common property theory and Critical Institutionalism	Analysis of dynamic interactions between purposeful and non-deliberate institutions	Household questionnaire, with purposive sampling; focus group discussions; informal discussions; observation; semi-structured interviews.	• The challenging of maintaining purposeful institutions, developed to 'govern' natural resources, may be explained by the influence of 'non-deliberate' institutions. • Mainstream institutionalism associated with the study of the commons should more explicitly recognise and analyse institutions beyond those crafted for natural resource governance.
7/Polycentricity within customary governance systems	Multiple levels and centres of authority within customary governance of marine ecosystems	Ethnography over 25 years, using data from open-ended interviews	• There are multiple centres of decision-making within the customary system and these affect the nature of rights that individuals within the same community. • Recognition and understanding of the existence of multiple centres and levels and how these interact should inform dialog and design around new approaches to natural resource governance.

Chapter/Approach	*Focus of analysis*	*Methodologies*	*Insights*
8/Political ecology	Power at the centre of analyses of resource governance and discourses	Interviews; analysis of multiple discourses	• Political ecology lens directs analysis towards inclusion and exclusion in resource access and decision-making. • Multiple forms of governmentality identified, associated with discourses and practices related to conservation.
9/Agency of local state actors	The agency of local state actors ('interface bureaucrats') in governing renewable natural resources	Interviews with key informants	• The state is not monolithic – different interests, resources and incentives, and operating in different contexts • Focus on interface bureaucrats highlights how state representatives and citizens interact on a daily basis as part of natural resource governance • Interface bureaucrats are linked to reproducing state authority – as regulator and service provider • Interface bureaucrats change policies and practices significantly during implementation
10/Analysis of legal and institutional pluralism	Investigates how diverse policies and programs that form part of forest governance operate in parallel and often result in new institutions being formed at the local level. Uses lens of legal and institutional pluralism.	Ethnography; repeat interviews over decades, tracing changes in policy, systems, projects and practices, and participant observation.	• The multiple institutions that result from internal legal pluralism do not necessarily interact with each other and there may be unclear or inadequate accountability mechanisms. • Argues for 'institutional consistency', whereby responsibility for natural resource governance at the decentralized level rests with elected local government.

(*Continued*)

Chapter/Approach	*Focus of analysis*	*Methodologies*	*Insights*
11/Development and governing through a compensation incentive payments scheme	Analysis of the potential for a compensation incentive payments (CIP) scheme to address grievances and shortcomings in past and current governance	Feasibility study including interviews, focus group discussions and village meetings. Stakeholder workshop convened to outline the framework for a CIP program	• Due to increasing complexity and dependencies, greater coordination and cooperation between actors is required • Co-design of policy instruments with affected stakeholders can result in a more locally-relevant and legitimate way forward • The three dimensions of governance, politics, polity and policy, should be considered together, with links between them acknowledged

arrangements, which suggests that policies, institutions and organizational arrangements in a governance regime reflect the balance of power among different constituents in a society'. This approach places power at the heart of the analysis of governance assessment frameworks and supports greater consideration of the political and economic context of governance. Political ecology, introduced and applied in Chapter 8, recognises the primacy of power in mediating people–environment relations, with Upton explaining that political ecology is 'a loose body of theory, approach or frame of enquiry…within which relations of power are central'. She identifies different roots of political ecology, drawing in the chapter on a post-structuralist perspective which focuses analysis on discourse and the 'politics of knowledge'. Building on this, she recognises the influence of Foucault's governmentality on how political ecology has engaged with 'the workings of power in environmental governance'. Through applying this approach to political ecology, Upton concludes that political ecology can investigate the 'discursive processes that underpin these more legible issues of rights of access and decision-making', thereby stressing the importance of understanding underlying discourses associated with natural resource governance as they influence governance approaches adopted, decision-making and outcomes.

In Chapter 3, social network analysis (SNA) is explained by Mancilla Garcia and Bodin as a way of analysing actors, their attributes and relational ties. They note that network approaches have been used in relation to the analysis of natural resource governance 'to study different forms of cooperation, collaboration and collective action'. Mancilla Garcia and Bodin explain that network analysis explores patterns of social relations and so can investigate which actors interact with whom, which actors are most important (or 'central') in a governance system and how information flows. SNA can enable clusters of actors to be identified and actors to link, or 'bridge', clusters of actors. This can then lead to analysis of why such clustering is happening and what the implications are for the practice and outcomes of natural resource governance, for example.

Within the chapters, there is significant attention to the role of institutions in natural resource governance, particularly informal institutions, also referred to as socially-embedded institutions (Chapter 5) and non-deliberate institutions (Chapter 6). This perspective on institutions reflects a more 'critical' turn in institutional analysis of natural resource governance, one that 'generally tends to reflect complex and unequal relations around natural resources management within and between the local, national and global levels' (Hall et al., 2014: 82). This is contrast to the 'mainstream' approach to institutional analysis and development (IAD) within commons literature, in which the 'right' institutions are crafted with the aim of delivering on sustainability (see Cleaver, 2012 and also Chapters 5, 6 and 9 of this volume). In Chapter 5, critical institutionalism is introduced as an approach that recognises that a diversity of institutions influences how a resource is used and who benefits, not just institutions crafted for that purpose. These institutions might include gender relations and norms, kinship, social norms and customary rules. Institutional bricolage takes this further by recognising that institutions do not operate in isolation but interact and shape each other and so institutions may change over time. Cleaver (2002) acknowledges inspiration from the work of Douglas (1973, 1987) in developing her perspective on institutional bricolage, noting that Douglas argued that institutions and decisions rarely result from rational choice.

Application of the concept of institutional bricolage within natural resource governance has particularly shown how externally introduced institutions, such as local structures associated with community forest management, have been rejected, modified or combined with other institutions resulting in different institutional arrangements in practice over time and space (de Koning, 2014). A critical institutionalist approach to analysing natural resource governance therefore requires the identification and analysis of a wide range of institutions within a natural resource governance context, preferably over time. Barnes takes a similar approach in Chapter 6 where she analyses the influence of 'non-deliberate' institutions on formal, or 'purposeful', institutions. She does this by drawing on both critical institutionalism and mainstream institutionalism, arguing that analyses informed by mainstream institutionalism should go further in turning to non-deliberate institutions to explain governance practice and outcomes. This involves explicit recognition and analysis of such institutions rather than a more limited acknowledgement of the influence of these institutions on purposeful institutions.

This critical perspective offers an alternative view to literature on natural resource governance informed by Ostrom's design principles and common property or pool theory, introduced in Chapter 1. Three chapters were, however, informed by this body of theory – Chapters 4, 6 and 7. In Chapter 4, Partelow introduces the social-ecological systems framework (SESF) and the background to it in the form of collective action theory and Ostrom's IAD framework. He explains how Ostrom and others' work on the commons has shown how collective action and cooperation between people does take place and can lead to sustainable use of common-pool resources (CPRs), countering claims encapsulated in Hardin's Tragedy of the Commons that people are rational utility maximisers and will endeavour to benefit individually from a resource, at the expense of sustainability. The SESF, as Partelow notes, brings a systems thinking perspective to natural resource governance, drawing on theory on social-ecological systems and providing a diagnostic tool to identify the natural resource use problem. The wide range of variables provided in the multiple tiers of the framework are then examined to understand how they affect the problem and how they interact with each other and with what consequences. This is the only approach in this volume that explicitly requires collection and analysis of data related to natural resource systems, though this is very much from an economic perspective.

Aswani in Chapter 7 takes an approach to analysis also informed by commons literature, that of polycentrism. He notes that this approach refers to 'the existence of a number of governance decision-making units that may be in a competitive and/or cooperative association with each other', using this perspective to analyse indigenous sea tenure. Polycentrism provides a lens to natural resource governance that recognises that there can be multiple centres of decision-making, providing often a more locally relevant and effective system of governance. Aswani applies this perspective in an original way, in the context of customary governance, and through this approach is able to explain the diversity of rights experienced by individuals within a community to marine and terrestrial resources.

Funder draws on several areas of theory and literature in Chapter 9 to investigate agency of local state actors, including literature on the anthropologies of the African state, public administration studies, organisation studies, development studies and anthropology. More specifically, the ideas of Lipsky's (2010) 'street level bureaucrats', Olivier de Sardan's (2015) 'interface bureaucrats' and Cleaver's (2012) critical institutionalism inform his approach to the analysis of how local state actors operate between the state and citizens, negotiating, bargaining and seeking compromise in a context of conflicting interests and priorities and constrained resources. The diversity and breadth of literature and theory provides a comprehensive framing for an analysis of the role, context and practice of local government officers operating within natural resource governance.

Plurality is a theme found in many chapters, particularly of institutions that shape and form part of a governance system. In Chapter 10, Faye combines a theoretical perspective on institutional pluralism with legal pluralism to explain the plurality of legal domains within state law. Drawing on public administration and anthropological conceptions of pluralism, he undertakes an analysis of the legal instruments introduced through decisions of different parts and levels of government that form the complex mosaic of natural resource governance within the context of decentralization. This perspective contributes explanation for the plurality of institutions that exist with the landscape of natural resource governance and how over time contradictory laws and interpretations of laws may result from uncoordinated reform.

Chapter 11 provides a further novel approach, analysing the nature of conservation incentive payments (CIPs) from the perspective of the politics, polity and policy nexus. This is in contrast to a focus on economic efficiency, which has received much attention in terms of determining the most efficient level of payment to achieve the desired behaviour changes and conservation outcomes. Pekor et al. refer to politics as 'the local socio-economic, political and cultural conditions', polity as 'the level of collaboration, engagement and fairness achieved in designing and implementing the scheme' and policy as 'the degree to which the implementing instrument is tailored to formal and informal institutions, laws and norms'. This nexus demonstrates the scale of the challenge in developing a policy instrument that aligns with a local context, will have legitimacy and be accepted, as well as have the potential to deliver on policy objectives.

The chapters present a diversity of approaches and literature on which to draw for analysis of natural resource governance. However, there are several common themes in these approaches and areas of literature. Analysis of institutions is particularly dominant, which is not surprising given the importance of institutions in influencing how decisions are made and people gain and maintain access to natural resources. It also reflects both the focus on institutions within literature on the commons, as reflected in Ostrom's own work, but also it reflects the diversity of approaches taken to analysing institutions. Even where slightly different perspectives are taken and literatures drawn on, similar conclusions are reached regarding the plurality and diversity of institutions involved in natural resource governance, the importance of socially-embedded institutions and that institutions interact with each other with implications for the institutions themselves as well as for social and ecological outcomes.

Further important points to note are the attention to power and politics within several of the analytical approaches, the focus on social relations within SNA and the link made to literature on public administration and policy instruments. There are then many questions that can be asked of natural resource governance, necessitating the use of different approaches, frameworks and areas of literature. This is encouraging as the diversity of questions and analytical approaches reflects the complexity and diversity of much natural resource governance.

Focus of the analysis

This section identifies and reflects on the focus of the analysis in each chapter, which ranges from frameworks for assessing natural resource governance to customary sea tenure. This illustrates how the focus of analysis can significantly vary within the broad area of natural resource governance.

In Chapter 2, Kashwan utilises insights from new institutional economics and theories of power to produce a comparative analysis of three established frameworks for assessing natural resource governance. He critiques the technical approaches these frameworks tend to take and develops a meta-framework that reflects the essential ingredients of a self-sustaining system of governance, especially given the 'context of entrenched political and economic inequalities'. Power then is central to his analysis of and response to the frameworks he reviews.

In Chapter 1, it was stated that governance in relation to renewable natural resources often involves users of a resource. This implies participation in structures and systems and collective action by multiple stakeholders. In Chapter 3, Mancilla Garcia and Bodin focus on whether participatory forums developed to encourage inclusion of, and coordination between, stakeholders are achieving those aims. Participation of affected stakeholders is considered essential in natural resource governance, though it is often challenging to achieve in a just, equitable and effective way, as explained in Chapter 1.

Chapters 4 and 5 focus on co-management as a form of governance where the users of a natural resource, fisheries in the case of both chapters, work with government and others in managing the resource. In Chapter 4, Partelow uses the SESF to examine variables influencing the governance of the Brazilian Extractive Reserve (RESEX) program in Bragança, particularly in terms of understanding which variables have affected how collective action has worked, with implications for fisheries management. In Chapter 5, the analysis focuses on how co-management structures and systems interact with socially-embedded institutions with implications for how co-management works in practice and for social, economic and ecosystem outcomes. Analysis through the lens of institutional bricolage provided explanation for the dominance of certain actors in co-management structures, why co-management struggled to reduce illegalities and the implications of the interaction between institutions for the state of the fisheries. In a similar vein, Barnes focuses her analysis in Chapter 6 on the formation and operation of a Forest Rights Committee in a village in Andhra Pradesh, India, formed under the Forest Rights Act of 2006, which gave forest dwellers the right to claim legal recognition of their rights to forests and forest resources. In particular, her analysis concerns why and how formal recognition of rights, and an auction system to sell bamboo set up as part of that recognition, was not maintained over time due to the influence of existing, informal, or 'non-deliberate', institutions. Each chapter recognises the complexity of the introduced governance system resulting from the number and range of factors that influence how the system is received, understood and works in practice, reflecting the historical, social, political and economic context within and beyond the governance system.

The importance of the historical and sociopolitical context is also emphasised in Chapter 7, where the focus of the analysis is on the multiple decision-making centres within a customary system of sea tenure. The focus of analysis on customary tenure reflects the prevalence and importance of such systems in many parts of the world and how these systems remain effective in natural resource governance. Aswani takes a unique approach to analysing a customary system by recognising many centres within it, including nested institutions, and how these work together to shape rights and resource access.

Upton takes a different approach in Chapter 8, focusing on an analysis of discourses that seek to explain environmental degradation and how those discourses shape and inform decisions about governance priorities and systems. The focus of the analysis concerns how discourses around conservation, carbon and mining in Mongolia portray and affect pastoralism. The analysis shows how herders are expected to adopt conservation-oriented behaviour externally deemed to be desirable, with implications for herder identity and livelihoods resulting from the consequent governance approach adopted.

Local government officers, or 'interface bureaucrats', are the focus of the analysis in Chapter 9, which Funder justifies through observing that the state is not monolithic; rather, there are multiple and diverse interests, levels and faces of the state. Such officers Funder lists as 'local managers, technical specialists and extension agents who implement state policies, laws and projects on the ground', with the term 'interface' giving emphasis to the relational dimension of their position between the state and the public. The focus of the analysis more specifically is on the everyday practice of these officers within environmental governance in terms of the context in which they work and what they draw on to get their work done. Such officers, Funder states, work with a 'high degree of discretion', reflecting local norms, bargaining and compromise, and within severe resource constraints.

The multi-layered order of law within the forestry sector is the focus of analysis in Chapter 10, associated with a plurality of institutional bodies. This focus on the legal arrangements provides explanation for the plurality of structures and systems discussed in other chapters and demonstrates the complexity of law resulting from multiple parts and levels of government over time. Faye examines the evolution of law within the forestry sector of Senegal over time, thus enabling an analysis of how legislation becomes multi-layered and presents challenges for coordinated and coherent natural resource governance. In addition, by focusing on the decentralized context of forest governance, Faye demonstrates the challenges resulting from developing institutions outside of the elected local government system for accountability and consistency.

Local context and acceptance are key in Chapter 11, where the potential for a new policy instrument, a CIP, is examined. The focus of the chapter is an analysis of the characteristics of a CIP scheme in relation to the politics, polity and policy nexus and a reflection on the process of co-design of the scheme. The analysis illustrates the interconnectedness of policy instruments with governance, with any boundaries between the two being blurred through the design, practice and perceptions of stakeholders, associating policy instruments with forms of governance and, in this case, seeking an alternative to top-down governance of a conservation area.

The focus of analysis can, then, be very varied, depending on the research questions and empirical situation. Whatever the focus of analysis, a common theme throughout the chapters is the importance of locating the analysis within the wider context – which may include an historical perspective and understanding of the relevant social, economic and political context. Such reference to the wider context recognises that any natural resource governance system is affected in terms of design, legitimacy, practice and outcomes by external as well as internal factors, as reflected in the SESF.

Methodologies

The contributions to this volume primarily used qualitative research designs and methods, particularly, but not exclusively, ethnography. The prevalence of ethnographic approaches in the contributions may reflect the desire to understand the complexity of governance in practice, which requires an immersion of the researcher in the empirical context, getting to know the people involved in the situation and those being observed or interviewed getting to know the researcher. The cases examined in the chapters have clearly shown that there are often multiple governance systems, structures, policies and institutions, making it hard to understand, interpret and disentangle for a researcher. The multiplicity and diversity of actors, interactions and institutions affect how governance is conducted in practice as opposed to how it is planned and/or intended to take place. This makes fieldwork essential and supports the use of qualitative research.

In many of the chapters, authors refer to collecting data from different types of stakeholders. This is critical as different people may have different views, perspectives and experiences of the same event or system. They may remember and interpret the same event, conversation and decision-making differently. An interpretive perspective is therefore often needed in research on natural resource governance, recognising that there is more than one 'truth' and so different perspectives, experiences and interpretations should be sought. This is vital when research is seeking to understand how governance is practised and how decisions were made, for example.

The most commonly cited method of data collection is the interview, though some chapters refer to repeat visits and data collected over time, suggesting familiarity with the setting and respondents that would enable more ethnographic interviewing to take place and a longitudinal study to be undertaken. In addition to ethnographic interviewing, Harrison (2018) identifies participant-observation and fieldnote writing as key data collection methods within ethnography. Participant-observation is particularly important for ethnography, as such an approach allows the researcher to move between 'insider' and 'outsider' status, though the lines between the two may be blurred. He goes on to note the importance of duration in the field setting so that people become accustomed to the presence of the researcher; of resemblance, both physical and social, between the researcher and people in the research setting, which may refer to languages spoken, age, gender and ethnicity; and, of the level of participation, which may be affected by multiple factors and have implications for the potential for observation.

Ethnography is a complex and challenging approach to take to research but many of the cases included in the volume show how valuable an immersive, longitudinal approach to data collection is in the context of natural resource governance.

There is only one example in the chapters that takes a quantitative approach to data collection and analysis and that is social network analysis. This approach involves mapping and ranking or valuing interactions between actors, whether through the importance or frequency of interactions. Mancilla Garcia and Bodin show in Chapter 3 how such data can be used to test hypotheses about how a network may operate and actors within a network behave.

The dominance of approaches and case studies that adopt a qualitative approach reflects both the use of case studies in many of the chapters and its dominance as an approach within research on natural resource governance (Poteete and Ostrom, 2008). The dominant use of case studies was explained by Poteete and Ostrom (2008) as reflecting the need to understand the contextual factors of each example of natural resource governance and for researchers to be familiar with each example. They note that there is a scarcity of large-N studies on collective action at the local level and suggest that this limits the 'accumulation of knowledge about natural resource management, with important implications for natural resource policy' (Poteete and Ostrom, 2008: 177). To address this scarcity, they recommend the adoption of the following strategies to generate comparative data for large sample sizes: development of meta-databases and use of partnerships and networks to generate data that can be used quantitatively within a large sample whilst retaining the grounding in qualitative data collection.

Partelow in Chapter 4 refers to the social-ecological systems framework as being at least in part developed to 'solve the comparability problem', that is, provide multiple social and ecological variables that can be used to inform data collection, coding and analysis across studies and cases. This has indeed been initiated under the SESMAD project, which refers to the development of a social-ecological systems meta-analysis database. This is an international collaboration that uses the SESF to enable common coding across studies, though in this case with a focus on large-scale ecosystems, with those involved in the initiative seeking to identify 'what variables and theories developed in the analysis of small- to medium-scale CPRs are also important in explaining success or failure in the long-term governance of largescale environmental systems?' (Cox, 2014: 266). How far the SESF presents a solution to the 'comparability problem' may be too early to tell and may be challenging to deliver on, particularly given the need for understanding of local context noted by Poteete and Ostrom (2008) which might encourage adaptation and modification of variables as well as differences in interpretation and measurement.

The chapters clearly support the use of qualitative research in analyses of natural resource governance, though the analytical approaches introduced and applied within the chapters are not necessarily representative of all the approaches that could or should be taken.

Insights

Within the column in Table 12.1 on 'insights' arising from the reflections on and applications of the approaches, frameworks and theories in the chapters, three clusters can be identified. The first cluster concerns the multiplicity and diversity of institutions within a governance system and multiple variables influencing governance systems, practice and outcomes. Given this multiplicity of institutions, several chapters highlighted the way in which institutions interact and with what consequences for governance and for livelihood and ecosystem outcomes. The second cluster then is concerned with interactions between institutions. The third cluster is 'governing in practice', recognising that it is almost universally the case that governing in practice is different to what was envisaged in policy and guidelines. These 'clusters' are interrelated and cannot be considered in isolation. However, categorising common and important emerging themes from the reflection and analysis enables points to be elaborated on and examples given. For example, whilst the third cluster of points, 'governance in practice' identifies reasons why governance generally appears and is experienced differently to how it was planned, the first two clusters on multiplicities of institutions and influencing variables and of interactions between institutions also offer explanations for this.

Multiplicity and diversity of institutions and influencing variables

Plurality and diversity of actors, systems, structures and institutions are common themes throughout the book. For example, Faye in Chapter 10 demonstrates how the introduction of multiple laws and structures associated with multiple donor-supported projects has led to a situation of legal pluralism within the 'single' realm of state law. This has created a complex system of governance which means that it is challenging to ensure that there is appropriate accountability and transparency of decision-making. From the analysis, Faye concludes that if governance approaches and projects were built into the existing local government system, there would be greater accountability to the local communities through elected councillors.

In Chapter 7, Aswani sets out the multiplicity and diversity of rights within a multi-level, or polycentric, system of customary tenure of marine resources. The case study demonstrates how rights of access to certain marine and terrestrial areas vary between individuals depending on their lineage. This confirms a long-held recognition that 'communities' cannot be portrayed or understood as homogenous, that is, having people with the same rights, interests, incentives and status (see Agrawal and Gibson, 1999). Instead, power differentials and differences in interests and assets affect whether, how and to what extent individuals and households have access to and benefit from natural resources. Aswani's analysis shows how a customary system can have multiple centres and how these create a complex system of rights which means that rights vary between individuals. This example illustrates the importance of recognising multiplicity within customary systems and of not making assumptions about individuals within a 'community'.

In addition to a focus on a plurality of institutions, levels and structures, through applying the SESF Partelow shows in Chapter 4 how there are likely to be many factors that influence the shape, practice and outcomes of governance, with interactions between factors further complicating governance systems and outcomes. He does, however, highlight a few as being particularly important in the case study, such as the lack of leadership and effective communication within the co-management system and insufficient incentives in the form of improved prices or compensation to motivate active participation by resource users in the governance system. This example and the framework itself demonstrate that it is unlikely to be sufficient to focus on one issue if reforming a governance system and awareness of how one component or factor may affect others is needed.

Interactions between institutions

Two of the chapters focus explicitly on how formal, or bureaucratic, and informal, or socially-embedded, institutions interact and with what consequences. In Chapter 5, the application of critical institutionalism demonstrates how socially-embedded institutions affect the participation of actors in fisheries co-management through power relations, social norms and social and economic relations. The introduction of co-management, as a bureaucratic institution, cannot be assumed to operate in isolation of socially-embedded institutions. These institutions interact and may be pieced together by stakeholders through a process of bricolage. The chapter demonstrated how bricolage affects the practice and outcomes of fisheries co-management, particularly by enabling more powerful actors to remain dominant at the local level and enabling illegalities to continue.

The approach and subsequent insights are similar to those found in Chapter 6 on community governance and institutional interfaces. In this chapter, Barnes initially introduces literature on factors identified by Ostrom (1990) and others as influencing the potential for institutions to sustainably govern CPRs. She refers to such institutions developed, or crafted, for the purpose of natural resource governance as 'purposeful' and contrasts these with institutions that influence access to and benefits from natural resources that were not designed for that purpose. These institutions she refers to as 'non-deliberate', examples of which are given as gender and caste norms. Barnes draws on critical institutionalism as a way of identifying socially-embedded institutions and their interaction with purposeful institutions. She argues that it is not enough for commons literature, or mainstream institutionalism, to recognise socially-embedded, or non-deliberate, institutions and suggest that they shape or inform purposeful institutions. Instead, she argues that non-deliberate institutions should be given greater attention in their own right and in how they interact with purposeful institutions. Barnes highlights examples of such interaction as social norms around male youth protecting the village interacting with the more formal monitoring of the forest and inter-village kinship ties preventing the imposition of sanctions on neighbouring villagers.

Key differences between Chapters 5 and 6 are the deeper embedding of the analysis in literature on mainstream institutionalism, related to Ostrom's design principles (Ostrom, 1990) and the IAD framework (Ostrom, 2011), in Chapter 6 and the explicit reference to and application of institutional bricolage in Chapter 5 as a way of envisioning and analysis of how institutions may be pieced together to form new or adapted institutions. The discussion and application of literature by Barnes is a useful reminder of the utility of literature and analysis that has stemmed from 'mainstream institutionalism' or common property theory.

Governance in practice

Several chapters pay attention to the lack of congruence between what is envisaged for the governance system and management of natural resources and what happens in practice. Analysis of how institutions interact discussed above can provide insight into why governance looks and is experienced differently to how it may be designed or envisaged, as well as lead to different outcomes. Further insights on the practice of governance can be generated by other analytical approaches.

In Chapter 4, Funder analyses how local government officers, referred to as 'interface bureaucrats', implement climate change policies in a way that is meaningful to them and the local communities. This illustrates that what happens at the local level may be different to what was planned or envisaged at the national level – there are likely to be differences in interests, perspectives and resources between different levels and parts of the state. His analysis illustrates the importance of negotiation, bargaining and compromise in the work of interface bureaucrats, reflecting their everyday interaction with users of the resource, and so having to live and work alongside people they may be sanctioning and introducing regulations that constrain their access to and benefits from natural resources. These approaches employed by interface bureaucrats also reflect the context of working with constrained resources, which limits their ability to move around and see what is happening on the ground. Interface bureaucrats are often having to find solutions that will to some degree satisfy multiple interests, from the ministry to local communities and local politicians.

Governance in design and practice may also be influenced by dominant discourses which privilege certain interests and assumptions above others. In Chapter 8, Upton identifies dominant discourses around conservation that inform policies and approaches that affect the identity and livelihoods of pastoralists in Mongolia. This suggests that explanation for why a certain approach to governance has been adopted may be found in underlying discourses.

Kashwan's development of a meta-framework in Chapter 2 focuses on the centrality of checks and balances for effective governance, informed by a critique of existing governance assessment frameworks that fail to ask important questions such as governance for what, by whom and under what conditions? The framework he develops and tests is informed by recognition of the importance of power in the practice of governance, which is apparent in the 'specification of goals', 'representation of stakeholders' and 'monitoring and enforcement' as well as the two dimensions of the meta-framework which explicitly refer to power.

In Chapter 3, Mancilla Garcia and Bodin demonstrate how SNA can be used to investigate the structures and practice of governance, by focusing on who people interact with, with what frequency and for what purpose. SNA can be used to identify who is influential in the practice of governance and which actors are marginalized and why. Such analysis can identify implications of the arrangements of social relations for natural resource management and livelihoods.

Finally, governing in practice is illustrated by the analysis of a potential policy instrument in Chapter 11, with CIPs described by Pekor et al. as a way to 'link economic interests of local people with conservation goals'. These policy instruments have been developed in recognition that local people often pay the financial cost of conservation and yet are often not compensated at all or adequately. CIPs aim to make conservation financially viable for those affected and aim to result in improved outcomes for the environment, in the case of Chapter 11, wildlife, and people. Such compensation, or incentives, is directly linked to actions and outcomes. Of course, as Pekor et al. note, a CIP scheme will not necessarily deliver on the desired outcomes. What is particularly noteworthy about the case they present is not so much the policy instrument itself but the importance of the co-design of the programme, developed through a multi-stage feasibility study feeding into a multi-stakeholder workshop and CIP Design Committee. So, whilst it is critical to consider incentives and compensation in relation to conservation, the process of identifying and developing appropriate mechanisms through which such incentives and compensation are delivered is also crucial for legitimacy and buy-in.

Conclusion

The comparison of the approaches, focus of analysis, methodologies and insights clearly support the portrayal of governance as complex, resulting from the multiplicities of actors, institutions and systems involved. These in turn are associated with the scale and multiple benefits from natural resources but also are a result of changes to policy and legislation over time, which may be associated with reform processes and multiple donor-funded projects. Such complexity presents challenges and opportunities. The complex landscape of governance can be challenging in that it is hard to fully comprehend and appreciate, accountability may be difficult to establish and enforce, and such multiplicity may result in potentially conflicting objectives and rules. However, such complexity also offers resource users ways of navigating through institutions to secure access to, and benefits from, natural resources (c.f. the so-called 'forum-shopping' (Benda-Beckmann, 1981) and institutional bricolage (c.f. Cleaver, 2012; and see Chapter 5)). It also enables participation of multiple actors operating at different levels, including different parts and levels of government.

It is clear as well from the chapters, and from the conclusion above regarding complexity, that institutions involved in natural resource governance system are not static. In particular, there were multiple examples of where new systems of governance, such as community forestry or fisheries co-management, have been reshaped and adapted over time and space, as people respond through their socially-embedded institutions to the introduction of bureaucratic institutions. This reminds practitioners and analysts that new institutional arrangements are not introduced into an institutional vacuum; there inevitably will be interactions with socially-embedded institutions, such as power and gender relations, that affect systems, structures, practice and outcomes.

This complexity is also a reminder to practitioners and analysts of the need for a deep understanding of the array of institutions within a situation, in line with many references in the chapters to the need for understanding of the local context and for institutions to be relevant and appropriate to the local context. This then implies that any reforms to a governance system must involve those who will be affected in a genuinely participatory way, as illustrated in Chapter 11 in the lengthy process of co-designing a policy instrument.

Developing sufficient understanding of a governance system is a challenging task, particularly given that there may be a diversity and multiplicity of rights and institutions within as well as between 'communities'. This reinforces the call by Agrawal and Gibson (1999) for those concerned with natural resource governance to recognise heterogeneity within place-based communities, with a diversity of interests and assets making generalisations about motivations and capacity inappropriate. In addition, Ojha et al. (2016) remind us of the interconnectedness of people within a geographical location to people of very many other geographical locations, implying that there will be multiple and diverse 'communities' rather than one 'community' and that people may be part of regional or global communities.

Given the complexity of natural resource governance, it is inevitable that different questions can be asked of governance systems and practice, and different methods used to collect and analyse data. The preference for ethnographic approaches in the chapters of the book reflects the dominance of case study research in studies on natural resource governance, which in turn reflects the need for a deep understanding and analysis of local context, in all its forms. The focus on qualitative research may also stem from the prevalent focus on what governance involves in practice and how actors get things done rather than on the theory and design of governance systems and approaches. The case studies were able to illustrate how people, particularly interface bureaucrats, may have to innovate and negotiate within natural resource governance, influenced by pressure from diverse interests and having access to very limited resources.

Of course, caution should be taken in drawing hard and fast conclusions from the chapters, given that the collection is not necessarily a representative sample of approaches to the analysis of natural resources governance nor are the case studies necessarily representative of similar sectors and contexts – other approaches could be taken and other case studies analysed. However, in each chapter, the authors have explained and provided rationale for their approach and have illustrated application in detail. Methodological details are provided in some chapters, with many pointing to sources where more information can be found on the methodology used.

Drawing on analysis informed by different perspectives and approaches in the chapters, clear conclusions about natural resource governance can be made beyond its complexity. These are that diversity and flexibility, with inclusivity, are key for legitimate and effective systems. There must be scope for locally specific solutions to natural resource governance and recognition that systems will change over time. This is not always desired or accepted by all stakeholders. Establishing a national system of protected area governance, of community forest management or fisheries co-management is attractive in terms of having a clear system, shared objectives and fair structures and rules. However, even within these, there could be greater scope for acknowledging and allowing local adaptations, which will happen anyway, given the evidence from several chapters in this volume that institutions interact and, as a consequence, are adapted over time.

This conclusion supports literature and practice that encourage adaptive management of social-ecological systems through adaptive governance. Adaptive governance requires systems to be in place that can respond to new information and change, being able to learn and experiment with management approaches (Folke et al. 2005). It implies an approach to governance that is collaborative and able to scale up to the challenge of large social ecological systems. However, as Chaffin et al. (2014: 9) observe, adaptive governance (AG) 'cannot be created by a unilateral action of government. AG is never the same in two places; it is messy and often develops organically within the context of a SES, but it can also be encouraged with an intervention aimed at boosting adaptive capacity. It exists across a spectrum, from top-down to bottom-up, from rigid to flexible, from global to local, but never exists at one extreme'. This supports the conclusion that within natural resource governance, there should be a diversity of systems and approaches, which are flexible in being able to respond quickly and effectively to new information, are locally specific and relevant, and genuinely inclusive.

Given that there is necessarily a multiplicity of institutions operating within many spaces of natural resource governance, more focus must be given to creating capacity and incentives for these institutions to support more sustainable and equitable use of natural resources rather than focusing on the formation of structures and rules. As additional stress and uncertainty results from climate change, greater understanding of, and ability to respond to, ecosystem change will become increasingly essential. The evidence from this volume suggests that locally specific and adaptive institutions should be supported, with the remit, capacity and incentives for sustainable and equitable natural resource governance.

References

Agrawal, A. and Gibson, C. C. (1999) 'Enchantment and disenchantment: The role of community in natural resource conservation', *World Development*, vol. 2, no. 4, pp. 629–649.

Benda-Beckmann, K. V. (1981) 'Forum Shopping and Shopping Forums - Dispute Settlement in a Minangkabau Village in West Sumatra', *Journal of Legal Pluralism and Unofficial Law*, vol. 19, pp. 117–159.

Chaffin, B. C., Gosnell, H. and Cosens, B. A. (2014) 'A decade of adaptive governance scholarship: Synthesis and future directions', *Ecology and Society* vol. 19, no. 3: 56. http://dx.doi.org/10.5751/ES-06824-190356

Cleaver, F. (2002) 'Reinventing institutions: Bricolage and the social embeddedness of natural resource management', *European Journal of Development Research*, vol. 14, no. 2, pp. 11–30.

Cleaver, F. (2012) *Development through bricolage: Rethinking institutions for natural management*. London, Earthscan.

Cox, M. (2014) 'Understanding large social-ecological systems: Introducing the SES-MAD project', *International Journal of the Commons*, vol. 8, no. 2, pp. 265–276.

de Koning, J. (2014) 'Unpredictable outcomes in forestry-governance institutions in practice', *Society and Natural Resources*, vol. 27, no. 4, pp. 358–371.

Douglas, M. (1973) *Rules and Meanings*. Penguin, Harmondsworth.

Douglas, M. (1987) *How institutions think*. Routledge and Kegan Paul, London.

Folke, C., Hahn, T., Olsson, P. and Norberg, J. (2005) 'Adaptive governance of social-ecological systems', *Annual Review of Environmental Resources*, vol. 30, pp. 441–473.

Hall, K., Cleaver, F., Franks, T. and Maganga, F. (2014). 'Capturing critical institutionalism: A synthesis of key themes and debates', *European Journal of Development Research*, vol. 26, no. 1, pp. 71–86.

Harrison, A. K. (2018) *Ethnography (Understanding Qualitative Research)*. Oxford University Press, New York.

Lipsky, M. (2010). *Street-level bureaucracy: Dilemmas of the individual in public services* (Expanded ed.). Russell Sage Foundation, New York.

Ojha, H. R., Ford, R., Keenan, R. J., Race, D., Vega, D. C., Baral, H. and Sapkota, P. (2016) 'Delocalizing communities: Changing forms of community engagement in natural resources governance', *World Development*, vol. 87, pp. 274–290.

Olivier de Sardan, J. P. (2015) 'Practical norms: Informal regulations within public bureaucracies (in Africa and beyond)', in Tom De Herdt and Jean-Pierre Olivier de Sardan (eds.) *Real governance and practical norms in Sub-Saharan Africa*. Routledge, London.

Ostrom, E. (1990) *Governing the Commons*. Cambridge University Press, New York.

Ostrom, E. (2011) 'Background on the institutional analysis and development framework', *Policy Studies Journal*, vol. 39, no. 1, pp. 7–27.

Poteete, A. R. and Ostrom, E. (2008). 'Fifteen years of empirical research on collective action in natural resource management: Struggling to build large-n databases based on qualitative research', *World Development*, vol. 36, no. 1, pp. 176–195.

INDEX

Note: **Bold** page numbers refer to tables; *italic* page numbers refer to figures and page numbers followed by "n" denote endnotes.

For Product Safety Concerns and Information please contact our EU
representative GPSR@taylorandfrancis.com
Taylor & Francis Verlag GmbH, Kaufingerstraße 24, 80331 München, Germany

www.ingramcontent.com/pod-product-compliance
Ingram Content Group UK Ltd.
Pitfield, Milton Keynes, MK11 3LW, UK
UKHW022316250926
475702UK00046B/874

* 9 7 8 0 3 6 7 1 4 6 6 9 6 *